David Schofield

INSTITUTE OF GEOLOGICAL SCIENCES
GEOLOGICAL SURVEY OF GREAT BRITAIN

The Geology of the Cross Fell area

(Explanation of 1:25 000 Geological Special Sheet comprising parts of Sheets NY 53, 62, 63, 64, 71, 72, 73)

I. C. BURGESS, B.Sc. and A. J. WADGE, M.A.

LONDON
HER MAJESTY'S STATIONERY OFFICE
1974

ISBN 0 11 880712 9

CONTENTS

LIST OF TEXT FIGURES

1/INTRODUCTION

The 1:25 000 map of the Cross Fell Inlier covers the western scarp of the northern Pennines and the lower ground to the west. The scarp is formed of flat-lying Carboniferous rocks. At its foot a variety of Lower Palaeozoic rocks is exposed in the Cross Fell Inlier; between Murton and Knock, the harder rocks in this tract form isolated sharp-pointed hills or 'pikes'. Farther west, the lower ground of the Vale of Eden is underlain by Permo-Triassic sediments.

The scarp is a major topographical feature of northern England and includes the highest Pennine peak, Cross Fell [6874 3433][1], 2930 ft (893 m) above OD. The principal watershed along the summit of the scarp rarely falls below 2000 ft (610 m) OD, and divides streams flowing west into the Eden from the eastward drainage of tributaries of the Tees and South Tyne. East of the watershed, the slopes are gentle, extensively covered by peat and cut by shallow valleys whereas to the west, short steep valleys are incised into the face of the scarp. The lowest parts of the Vale of Eden lie below 400 ft (122 m) OD.

The rich soils of the Vale sustain both arable and livestock farming, but the Pennine slopes provide only rough grazing for sheep and horses and extensive areas are used as grouse-moor. In recent years the Moor House Nature Reserve has been established around Great Dun Fell and Cross Fell, whilst the area of moorland around Roman Fell lies within the British Army General Training Area at Warcop.

Mining is no longer as widespread in the district as it was during the 18th and 19th centuries. Major workings are now restricted to the gypsum-anhydrite mines in the Eden Shales near Kirkby Thore, but both lead and zinc were formerly

[1]Such numbers are National Grid references within the 100-km square NY.

worked on the Pennines. They occurred in veins particularly within the limestones (Dunham 1948, pp. 123–41). In places, the adits, opencuts and tips of these old mines can still be seen, and specimens of galena or sphalerite collected. Though many old levels are still open they are generally in a highly dangerous condition and should not be entered without expert guidance. In seeking new veins, the old miners cut many hushes on the fellside by damming small streams and repeatedly flushing the slope below to expose the bed-rock. Several hushes remain as dry, straight gullies; the deepest is Dun Fell Hush on the south-east slope of Great Dun Fell. The most extensive workings of vein mineralization in the district, at Silverband Mine on Great Dun Fell, were worked for barytes until 1963 and are now being re-opened as a result of the increased demand for this mineral as a drilling mud in offshore exploration.

Small-scale quarrying was once widespread in the district but has now ceased. Roadstones were worked from massive tuffs on Knock Pike, and from tholeiite dykes at Deep Slack near Melmerby. Elsewhere, many small quarries in sandstones and limestones provided local building- or walling-stone, and some limestones were also worked for lime. Small diggings for coal also occur.

In addition to being an area of outstanding natural beauty, the Cross Fell district is of great scientific interest, as recognized by the establishment of the Moor House Nature Reserve. It has long been one of the classical areas of British geology, containing a variety of strata and a complexity of structure seldom seen within the confines of such a small area. Since the earliest description by Buckland (1817) it has been visited by generations of geologists, both amateur and professional, and been the subject of much detailed work. Two publications in particular, by Nicholson and Marr (1891) and Shotton (1935), laid the foundations for this resurvey.

Geological History

The succession of rocks occurring in the district is shown on the vertical sections accompanying the map and summarized in Table 1. The solid rocks range in age from Ordovician (about 500 million years old) to Triassic (220 m.y.). On the lower ground of the Vale of Eden they are largely obscured by superficial deposits of various types laid down during Pleistocene and Recent times (probably mainly within the last 30 000 years) whereas parts of the high ground are mantled in periglacial (solifluction) deposits overlain by a variable thickness of peat. Several of the rock groups are separated by unconformities representing periods of time during which earth movements and erosion took place.

PERMIAN AND TRIASSIC	
St Bees Sandstone	250m
Eden Shales	180m
Penrith Sandstone	100-275m
unconformity	
UPPER CARBONIFEROUS	
Millstone Grit Series	205m
LOWER CARBONIFEROUS	
Alston Group	335-375m
Orton Group	30-125m
Basement Beds	65-300m
unconformity	
? DEVONIAN	
Polygenetic Conglomerate	0-100m
unconformity	
SILURIAN	
Brathay Flags	75m
Browgill Beds	85m
Skelgill Beds	42m
ORDOVICIAN	
Coniston Limestone Group (Dufton and Swindale Shales)	440m
unconformity	
Borrowdale Volcanic Group	1000m
fault	
Skiddaw Group (Kirkland Formation)	?
fault	
Skiddaw Group (Murton Formation)	?

Table 1. *Summary, with thicknesses, of strata present in the Cross Fell district*

The Lower Palaeozoic history begins in early Ordovician (Arenig) times, when the Cross Fell district lay within a rapidly subsiding geosyncline and a great thickness of greywacke grits and shales accumulated (Murton Formation). These beds became finer grained towards the top, and were followed in Llanvirn times by graptolitic mudstones, interbedded in their lower part with submarine andesitic and spilitic tuffs and rare lavas (Kirkland Formation). The rocks of these two formations, together included within the Skiddaw Group, were subsequently folded, uplifted and deeply eroded. On the eroded surface, there accumulated up to 1000 m of subaerial acid tuffs and volcanic sandstones, the lateral equivalents of the much thicker Borrowdale Volcanic Group of the Lake District. After a further period of folding and erosion, this volcanic landmass was gradually submerged by a shallow shelf sea in which the limestones and shales of the upper Ordovician Coniston Limestone Group were formed. The beginning of the Silurian (Llandovery) was marked by an increase in water depth and a return to the deposition of graptolitic mudstones (Skelgill Shales). These were followed by mainly barren mudstones, with thin graptolitic bands (Browgill Beds). During Wenlock times, the main deposits were again graptolitic siltstones and mudstones (Brathay Flags), the youngest strata of Lower Palaeozoic age preserved. The greywacke grits and shales characterizing the Ludlow Series in the Lake District are not exposed, but are presumed to have been laid down over the whole area, before the onset of the end-Silurian phase of the Caledonian orogeny, during which all the Lower Palaeozoic rocks were folded and cleaved.

Towards the end of this orogeny, the Weardale Granite batholith was emplaced in the area to the east. This intrusion is a factor of fundamental importance in the subsequent history of the north of England, forming the core of the rigid, low-density Alston Block (Bott 1967, pl. 6) and making it a positive area, almost continuously influencing local sedimentation, structure and topography through to the present day. The Cross Fell Inlier lies on the fracture zone marking the western edge of the Block—the Pennine Line. This structure has been active intermittently since end-Silurian times, with major movements occurring during the Caledonian, Armorican (late-Carboniferous) and Alpine (Tertiary) orogenies (Bott 1974).

During the Devonian period, the district was uplifted and deeply eroded. Deposits of this age are restricted to isolated fans of coarse detritus of Old Red Sandstone facies (Polygenetic Conglomerate). In Lower Carboniferous times, as the relief was gradually reduced, the hollows were filled with fluviatile sandstones and conglomerates of westerly derivation (Basement Beds) which probably formed part of a fluvio-deltaic complex built out into a Tournaisian sea to the south. Throughout the deposition of the succeeding Viséan Series, subsidence and sedimentation were evenly balanced. The depositional interface lay close to mean sea level, sometimes above, sometimes below, and there was an alternation of marine and deltaic deposition. In the lower part of the sequence (Orton Group and lower Alston Group) the environment was dominantly marine and the main deposits were shallow-water limestones. The overlying strata (upper Alston Group) display a rhythmic alternation of limestone with shales and sandstones, the 'Yoredale' sequence. In the succeeding Upper Carboniferous (Millstone Grit or Namurian Series), deltaic deposits predominated, consisting mainly of sandstones and shales, with only thin, though widespread, limestones.

Following the deposition of the Millstone Grit strata, there is another local gap in the sedimentary record. No strata of Coal Measures (Westphalian) age are preserved, though they are assumed to have been present over the whole area. During this period, the Carboniferous rocks were subjected to the stresses of the Armorican orogeny; faulting and folding took place along the Pennine Line, and the quartz-dolerite Whin Sill was intruded into the sedimentary rocks of the Alston Block in the late Westphalian or early Stephanian.

At a late stage in these movements, subsidence of the Vale of Eden area combined with a down-west movement on the Pennine Line to form an intermontane basin in which red dune-bedded desert sandstones and wadi breccias (Penrith Sandstone) of Lower Permian age accumulated. Mainly arid conditions persisted through the Upper Permian, though the nearby Zechstein sea occasionally flooded the valley, and led to dolomitic limestone deposition, and to the establishment of coastal sabkhas in which beds of gypsum-anhydrite accumulated at intervals in a mainly continental siltstone and sandstone sequence (Eden Shales). Throughout the Permian the western edge of the Alston Block formed an escarpment, which was gradually buried by the desert sediments, and was probably finally covered by the fluviatile sandstones of the succeeding Permo-Triassic St Bees Sandstone.

There is another long gap between the deposition of the St Bees Sandstone and of the Superficial deposits of Pleistocene times. Late in this interval, the marginal effects of the Alpine orogeny caused further down-west movement of the Pennine Line, and formed a fault scarp, the degraded edge of which is the present Cross Fell escarpment. During the Pleistocene period, the low ground of the district was invaded by ice from the Lake District, and the major valleys of the escarpment nourished small glaciers flowing both east and west. Each source provided its own identifiable boulder clay, whilst morainic material accumulated round the ice margins. The higher parts of the escarpment, outside the ice-sheet, were subjected to extreme frost conditions, leading to solifluction and cambering on most slopes. As the ice-sheet wasted away, glacial meltwater cut a network of deep glacial channels on the lower ground, and laid down scattered areas of sand and gravel. These glacial deposits provide positive evidence for only one period of glaciation, believed to be of late Devensian age, though earlier glacial episodes may have occurred.

2/ORDOVICIAN

Outcrops of Ordovician rocks form the main part of the Cross Fell Inlier. The principal lithostratigraphic subdivisions, and their classification and zonation, are shown in Fig. 1.

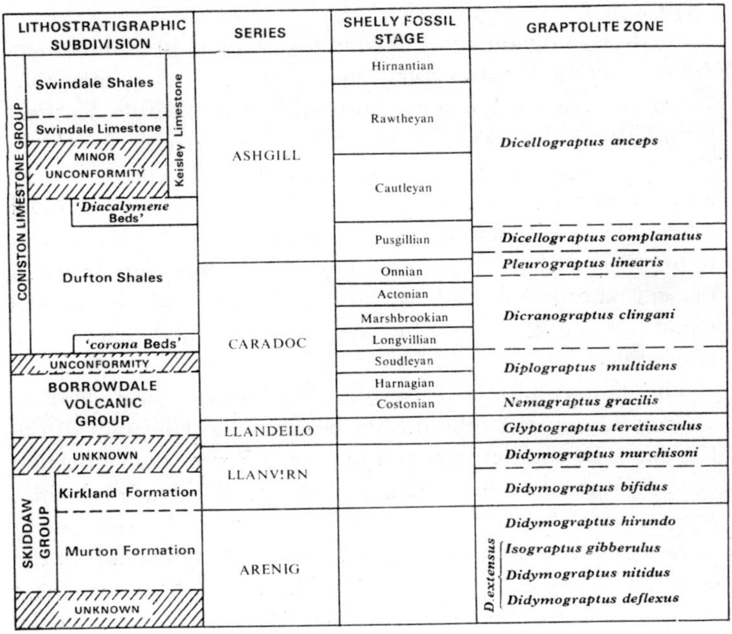

Fig. 1. *The Ordovician stratigraphy of the Cross Fell Inlier*

Skiddaw Group

The group consists mainly of argillaceous sediments, interbedded with subordinate sandstones in its lower part (Murton Formation) and with volcanics higher in the succession (Kirkland Formation). The base of the group is not exposed and its total thickness is unknown, although at least several thousand metres of beds are present.

The biostratigraphy of the Skiddaw Group has long been based upon graptolites. The Lake District outcrops yield faunas from the Arenig *Didymograptus extensus* and *D. hirundo* zones and the Llanvirn *D. bifidus* and *D. murchisoni* zones. Within the inlier, apart from one locality in the Murton Formation which yielded a possible *D. hirundo* Zone fauna (Fig. 2; Roman Fell, c), only *D. bifidus* faunas are present. These come from the Kirkland Formation and can be subdivided into lower and upper *bifidus* faunas. The latter is characterized particularly by abundant *Nicholsonograptus fasciculatus* and *Cryptograptus tricornis schaeferi* (Skevington 1970).

Analysis of the microfossils from the group (Lister and others 1969) has established a zonation apparently as precise as that based on graptolites, and more widely applicable to non-graptolitic facies. Within the inlier, two Arenig and two Llanvirn assemblages have been distinguished. The acritarchs and chitinozoa are particularly plentiful in the Llanvirn rocks, where the two assemblages correspond with the lower and upper *bifidus* Zone graptolite faunas. Within the limits of present knowledge, the Murton and Kirkland formations equate broadly with the Arenig and Llanvirn series respectively.

The faunas obtained during the resurvey provide new information on the relationship between the volcanics of the Kirkland Formation and the overlying Borrowdale Volcanic Group. Previous views, either that the Kirkland volcanics represented the initial stages of Borrowdale volcanicity (Green 1919), or that they were the seaward equivalents of the Lakeland Borrowdale sequence (Shotton 1935), led these workers to place the volcanic beds ("Milburn Group") at the top of the Skiddaw succession. It then followed that any associated non-volcanic sections ("Ellergill Group") were necessarily older. The graptolites and microfossils now show that the non-volcanic mudstones in Eller Gill are younger than any part of the Kirkland volcanic sequence, and similar non-volcanic sequences overlie the correlatives of the Kirkland volcanics in the eastern Lake District (Wadge 1972). Accordingly, it is now considered that the Kirkland volcanics represent a separate phase of activity, considerably predating the Borrowdale volcanicity.

Murton Formation

The formation consists largely of siltstones interbedded with subordinate mudstones and sandstones; the rocks cropping out at the northern end of the inlier contain a much higher proportion of sand-grade clastics than those farther south. A broad comparison with the Skiddaw Group sequence in the Lake District (Eastwood and others 1968, fig. 4) appears to suggest that these northern rocks therefore lie lower in the succession than those in the south, but this is not at present firmly established. The rocks all belong to a greywacke suite and were laid down under geosynclinal conditions.

In the southern part of the inlier pale to dark grey striped siltstones are interbedded with pale grey sandstones 5 to 50 mm thick. The rocks are generally highly cleaved. In hand-specimen the bedding is usually indicated either by faint colour-banding or by variations in grain-size. Thus it is typically picked out by paler-coloured sandy bands within the grey siltstones, although bands of quartz recrystallized along tectonic foliations may sometimes appear similar. Where cleavage and bedding coincide the resulting surfaces are commonly lustrous due to the development of secondary chlorite. Extensive quartz-veining is common in places, as on Brownber [7062 2748]. The rocks are best exposed on the ridge north of Keisley Beck [7220 2420] and on the south side of Brownber [7050 2720], whilst stream-sections are found in Murton Beck [7380 2235], Keisley Beck [7230 2366], Swindale Beck [6956 2854] and Sink Beck [6960 2890]. In those exposures north of High Cup Gill near the Carboniferous unconformity the slates are purple to a depth of several metres.

In the northern part of the inlier, grey siltstones are commonly interbedded with greyish green fine-grained greywacke sandstones which are generally about 5 to 50 cm thick, but occasionally exceed 5 m. They consist of poorly sorted angular to subangular sand which includes a wide variety of lithic clasts set in a clay matrix. Graded bedding can be seen in the coarser beds [6434 3750], and convoluted bedding is common in the thicker units [6435 3628]. Similar textural characteristics in the associated finer-grained beds show them to be greywacke-siltstones. Throughout this area the dominant surface is the bedding; cleavages are generally weak or absent and few minor folds are present. The beds strike generally east-north-east, and two major folds with this Caledonoid trend can be distinguished (Fig. 15), a syncline just north of Cuns Fell and an anticline farther south on Thack Moor. On the northern limb of the syncline at least 1000 m of beds are exposed, with little variation in lithology.

Locally, the finer-grained beds in the formation show chloritic spotting due to low-grade thermal metamorphism, particularly in the eastern part of this northern outcrop. Some occurrences can be related to nearby intrusions, such as the Cuns Fell dolerite [6484 3661], but others [6424 3747 and 6303 3671] may lie within the thermal aureole of the Weardale Granite.

The formation is exposed in Melmerby Beck [e.g. 6317 3688] and its tributaries, especially Dry Sike and Hungrigg Sike, although the latter yield largely strike-sections. Towards the head of Dry Sike [6446 3753], the beds are purple and hematitic, with poorly-defined bedding in a weathered zone, 30 m deep, below the sub-Carboniferous unconformity. Farther south there are many small exposures in the deeply-channelled ground north of Catterpallot Hill and also in Dale Beck [6360 3591 to 6368 3592], whilst around Deep Slack [6429 3564] several crags are formed of baked sediments adjacent to a large dolerite intrusion.

Kirkland Formation

The mudstones of the formation are interbedded with many thick bands of tuff and a few thin lavas giving a lithological sequence quite distinct from that of the earlier wholly sedimentary Murton Formation. It seems likely that the depositional environment was a north-easterly trending island arc, with several volcanic centres separated one from another by shallow seas, in which much of the tuffaceous debris was reworked. Graptolites are relatively common not only in those parts consisting largely of mudstone but also in thin shaly bands within mainly volcanic sections. The faunas are detailed in Fig. 2.

The base of the formation is not exposed, and the total thickness of strata present cannot accurately be estimated. An unbroken succession, about 500 m thick, is present around Wythwaite Top and at least 800 to 1000 m of beds crop out

SPECIES	Ashlock Sike a	Ashlock Sike b	Wythwaite a	Wythwaite b	Ellergill a	Ellergill b	Mudgill Sike	Burney Hill	Milburn Beck a	Milburn Beck b	Milburn Beck c	Milburn Beck d	Milburn Beck e	Milburn Beck f	Milburn Beck g	Milburn Beck h	Swindale Beck	Roman Fell a	Roman Fell b	Roman Fell c
Amplexograptus confertus (Lapworth)						×				×	?	×	×							
A. sp.		?			?															
Aulograptus cucullus (Bulman)				×	?												×			
Azygograptus coelebs Lapworth																?				
Climacograptus tailbertensis Skevington																		?		
Cryptograptus tricornis schaeferi Lapworth					×			×		?				×			×	×		
C. sp.					×										?					
Dichograptids indet.															×	?				
Didymograptus acutus Ekström																	×			
D. affinis Nicholson					?						×						×			
D. ex gr. *affinis* Elles and Wood										×	×									×
D. artus Elles and Wood					×	×	?										?		×	
D. bifidus (Hall)						aff.													aff.	
D. dubitatus Harris and Thomas	cf.																			
D. euodus Lapworth											?									
D. hirundo Salter																				cf.
D. pakrianus Jaanusson	cf.				×													cf.		
D. sp. (extensiform type)		×																×		
D sp. (pendent type)	×		×	×	?		×													
Diplograptus ellesi Bulman						×														
Diplograptids indet.										×				×		?				
Glyptograptus dentatus Brongniart					×			×		×				×			?			
G sp	?	?			?		?													×
Nicholsonograptus fasciculatus (Nicholson)					×	×			×									×		
Phyllograptus angustifolius (Hall)							×													
Pseudoclimacograptus angulatus (Bulman)							sl.													
P. cumbrensis (Bulman)		?																		
Trichograptus fragilis (Nicholson)											×									

LOCALITIES

Ashlock Sike	(a)	[6460 3492]	*Mudstones in north bank of stream*
	(b)	[6467 3491]	*" " " " " "*
Wythwaite	(a)	[6633 3276]	*Mudstones in west bank of channel—locality 'c' of Shotton (1935)*
	(b)	[6617 3273]	*Mudstone band in thick tuffs—locality 'a' of Shotton (1935)*
Eller Gill	(a)	[6768 3117]	*Mudstones in south bank of stream*
	(b)	[6757 3120]	*" " " " " "*
Mudgill Sike		[6765 3009]	*Mudstones interbedded with tuffs in track-side*
Burney Hill		[6862 3090]	*Mudstones in north bank of small stream*
Milburn Beck	(a)	[6787 2922]	*Mudstones in north-west bank of beck*
	(b)	[6791 2930]	*" " " " " "*
	(c)	[6797 2936]	*" " " " " "*
	(d)	[6838 2956]	*Mudstones in south bank of beck*
	(e)	[6855 2965]	*Mudstones in south bank of tributary*
	(f)	[6865 2963]	*" " " " " "*
	(g)	[6859 2982]	*Siltstones in east bank of beck*
	(h)	[6865 2992]	*Mudstone band in tuffs on east bank of beck*
Swindale Beck, Knock		[6931 2856]	*Mudstones interbedded with tuffs near junction with Sink Beck*
Roman Fell	(a)	[7491 2102]	*Mudstones in hillside exposure*
	(b)	[7492 2108]	*" " " "*
	(c)	[7507 2126]	*Siltstones in roadside exposure*

Fig. 2. *Lower Ordovician graptolites from the Cross Fell Inlier* (Identifications by Prof. D. Skevington)

on the well-exposed ground nearby, but the full succession is probably much thicker.

The proportion of tuff within the exposed parts of the succession, on Wythwaite Top, Burney Hill, Flagdaw, and near Keisley and Roman Fell, is comparatively high, but mudstones predominate in the intervening, low-lying areas for example around Eller Gill, Milburn Beck and Hilton Beck. No systematic thickening of the volcanic rocks can be distinguished.

The tuffs are generally massively bedded and commonly form low crags. They are largely andesitic, though spilitic[1] tuffs locally predominate as near Keisley. The tuffs are usually fine- to medium-grained, although in places they pass rapidly into much coarser material, either sharply across a bedding-plane or, more rarely, within a unit showing graded bedding. Most of the tuffs are heterolithic, and many are also crystal- and pumice-tuffs; lava lapilli are common near the lava outcrop at Wythwaite and also on Flagdaw and near Keisley. In a few tuffs the glass shards are tightly packed and appear to be partly welded. Many of the tuffs grade into volcanic sandstones in their uppermost parts where debris, including thoroughly broken shards and partly rounded clasts, has been reworked by the sea before final deposition.

The only exposures of lava within the formation are two spilite flows cropping out at Wythwaite Hole (Fig. 3) and an andesite seen within a sequence of tuffs west of Roman Fell.

The northernmost section in the formation lies in the headwaters of Ashlock Sike [6470 3492 to 6455 3487] where grey brown-weathering mudstone with thin ironstone bands is strongly folded but only weakly cleaved. Graptolites (Fig. 2) can be collected low down on the northern bank of the stream [6460 3492 and 6467 3491]. In several large scars in the southern bank of Ardale Beck [6468 3436 to 6504 3427], thin bands of greyish green reworked tuff in grey silty mudstones are exposed, generally with a steep dip eastwards. The thickest tuff hereabouts, 14 to 18 m thick, is coarse-grained and lies at the eastern end of this section. The low drift-covered ground immediately to the south appears to be underlain largely by mudstones, as the volcanic outcrops

---

[1]*Spilite, spilitic*. These terms are used here in the sense of basalt variably enriched in soda, the mechanism being unspecified.

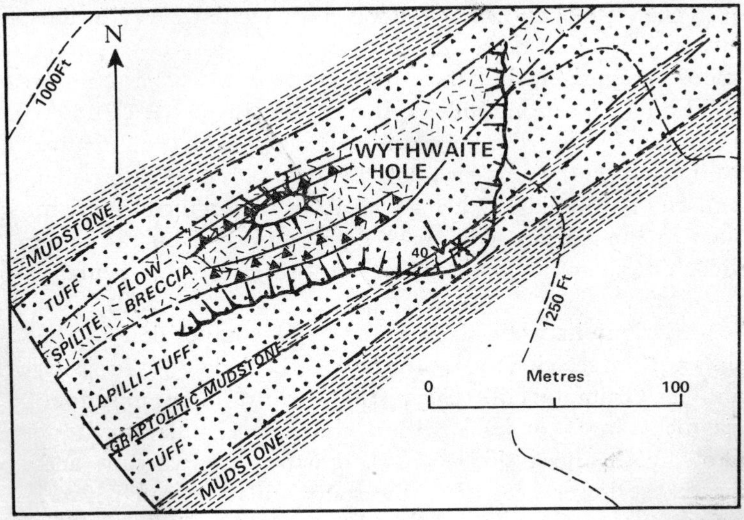

Fig. 3. *Sketch-map of the volcanic rocks at Wythwaite Hole* [662 327]

around Wythwaite Top produce much higher, craggy topography.

In the crags at Wythwaite Hole [6611 3278 to 6618 3269] shown in Fig. 3 the section reads:

	m
Tuff, greyish green weathering to pale grey, massive medium-grained crystal-lithic, siliceous . .	7.32
Mudstone, grey silty; partly transgressed by overlying tuff; *Aulograptus cucullus* . . .	0.61 to 0.76
Siltstone, grey, finely banded, volcanic . . .	0.61
Tuff, grey, fine-grained crystal-lithic, silicified .	1.83
Tuff, green, fine-grained; pale green lapilli up to 1 cm across	4.57
Spilite, greenish grey, aphanitic, flow-brecciated, siliceous	2.74
Spilite, greenish grey, aphanitic, siliceous . .	3.05
Spilite, greenish grey, aphanitic, amygdaloidal finely flow-brecciated?, siliceous . . .	2.44
Spilite, greenish grey, aphanitic, siliceous . .	5.18

The flow-brecciation of the upper part of the spilite in this section, its petrography and gradation into the overlying lapilli-tuff, suggest that it is a lava rather than a sill. Several spilitic tuffs, some containing spilitic lapilli, crop out nearby [e.g. 6645 3268] and this close association with the Wythwaite lava suggests that both were derived directly from a common spilitic magma, rather than that the lava was a basalt albitized after extrusion. Indeed, it has been suggested (Hudson 1937, p.398) that the same spilitic magma produced intrusive equivalents in the soda-rich dolerites of Cuns Fell, Catterpallot Hill and Baron Side. Other spilitic rocks within the formation, such as the Keisley tuffs, may have resulted from the same volcanic phase as the Wythwaite tuffs, but this remains uncertain.

Variations in grain-size, texture and composition of the tuffs are best seen in crags at the head of Littledale [6710 3210] and on Grumply Hill [6695 3181], whilst large mudstone fragments are common within the coarse tuffs north of Crowdundle Beck [6782 3180]. Farther downstream and across the lower ground to the south tuffs are much rarer, being best seen in the eastern bank of Mudgill Sike. The associated graptolitic mudstones are sharply folded in the south-west bank of Eller Gill. Graptolites are most plentiful at the north-western end of this classical locality [6768 3117 and 6757 3120].

Farther south, the tuffs exposed on Burney Hill are more heterolithic than those previously described, and unusually contain many clasts of acid lava of unknown source. Secondary silicification is common, giving quartz-veins on some crags [6861 3015].

The hill of Flagdaw consists almost entirely of andesitic tuffs and volcanic sandstones, commonly with large (up to 10 cm) fragments of black mudstone. The volcanics appear to lie on the crest of a dome, and younger beds come on to the north-west, north-east and south-east. The higher strata, exposed in Swindale Beck, Sink Beck and in a glacial drainage channel south of the Dun Fell road [6928 2920] are andesitic tuffs with interbedded bands of black mudstone. Similar mudstones are seen east of Dufton Pike between the Dufton Pike and Rundale faults. North of Keisley and north-east of Studgill, a large area of volcanics is assigned to this formation. The rocks are largely andesitic lapilli-tuffs with beds of volcanic sandstone. In places they contain abundant lapilli of spilitic lava together with conspicuous fragments of pink jasper. They are interbedded with mudstones which are exposed at the north end of Studgill [714 242]. South of Murton Pike much of the low-lying ground south of the Swindale Beck Fault is underlain by mudstones with sporadic bands of tuff. They are exposed in Murton Beck [732 221 to 733 222], in the sides of the deep glacial channels to the south, and on the south bank of Hilton Beck. West of Roman Fell tuffs lithologically similar to those of Keisley are exposed on the high ground east of The Seat. The area is faulted on all sides, and the exposures include vesicular, flow-banded andesite. The field relations of this rock are not clear, and though shown on the map as a lava, it may be an intrusion.

Borrowdale Volcanic Group

The Borrowdale Volcanic Group, represented in the Lake District by a thick sequence of lavas and tuffs, forming many craggy mountains, is confined in the inlier to several small, mainly fault-bounded blocks (Hudson 1937; Shotton 1935). The massive volcanic rocks tend to stand above the surrounding softer beds as steep-sided hills or 'pikes'. Two distinct sequences are exposed, respectively in the south and north of the inlier. The southern sequence, cropping out between Roman Fell and Milburn, can be matched with the southern outcrop of the Borrowdale Volcanic Group in the Lake District. The northern sequence, restricted to two small fault blocks east of Melmerby, correlates more closely with the volcanic rocks of the northern Lake District. The stratigraphical relationship of the Borrowdale Volcanic Group to the Skiddaw Group in the inlier cannot be demonstrated, the junctions in every instance being faulted.

Southern Sequence

Acid volcanic rocks crop out sporadically along the southern half of the inlier. They form the conspicuous pikes behind Knock and Dufton, and are also seen north of Keisley and beneath the Carboniferous conglomerates on Roman Fell. The rocks are assigned to three lithologically distinct formations totalling over 1100 m in thickness. The sequence is:

Harthwaite Tuff Formation:	Acid lapilli-tuffs, volcanic sandstones and siltstones . .	about 300 m
Knock Pike Tuff Formation:	Rhyolite and acid ash-flow tuffs	500 m
Studgill Tuff Formation:	Lapilli-tuffs and volcanic sandstones	300 m

These formation names are not indicated on the Cross Fell map. The Knock Pike Formation is shown as "Rhyolite and Acid Ash-flow Tuffs" (symbol R) and the Studgill and Harthwaite Tuff formations are included in "Acid Tuffs and Volcanic Sandstones" (symbol Z^R).

Studgill Tuff Formation

These beds crop out in Studgill and on the north-eastern face of Dufton Pike and may underlie wholly drift-covered ground north of Milburn Beck. They are generally poorly exposed, but appear to be mainly lapilli-tuffs and volcanic sandstones, possibly as much as 300 m thick on Dufton Pike.

Knock Pike Tuff Formation

The hard, resistant rocks of this formation are widely exposed, cropping out in Milburn Beck, on Knock Pike, Dufton Pike, Dod Hill, Gregory, Keisley Bank, The Seat, and on the northern and western slopes of Roman Fell. They consist mainly of rhyolitic ash-flow tuffs (ignimbrites), fine-grained and compact in hand specimen, and with a blotchy or streaky appearance, varying with the degree of compaction and welding and the abundance of volcanic clasts. The latter are more numerous in the upper part of the formation. As a result of weathering and oxidation, both in Permian and in more recent times, the tuffs are multi-coloured, ranging from pale pink and pale and dark green, through yellow and cream to white. The rocks consist of devitrified glass shards with a variable proportion of sodic feldspar crystals, pumice and acid tuff clasts except where recrystallization has destroyed all trace of original texture.

They are best seen in the large roadstone quarry on the north-east side of Knock Pike [6870 2850], where the most densely welded part of a thick ash-flow unit is exposed, lying near the base of the formation. The pale to dark green, fine-grained rock has a strongly developed streaky (parataxitic) texture, resulting from intense compaction of the included pumice fragments. The tuffs are closely jointed parallel to the parataxitic plane, and also display a rude columnar jointing perpendicular to this plane, extending from top to bottom of the quarry. Just to the south, in Swindale Beck [6892 2809 to 6891 2786], similar rocks are overlain by less densely welded tuffs with a blotchy appearance and well-developed lamination, containing numerous clasts of pumice and acid tuff (eutaxitic texture). The succeeding beds are poorly exposed, but appear to be mainly unwelded lapilli-tuffs, which are in turn overlain further downstream by eutaxitic welded tuffs [6888 2783 to 6884 2777].

In Milburn Beck, especially in crags on the western bank [6780 2872 to 6774 2856], the tuffs are deep red, possibly as a result of early Permian oxidation.

There are few other good sections, though the tuffs may be examined in hillside crags on Dufton Pike and in the fault blocks north of Keisley. On Roman Fell eutaxitic welded tuffs crop out in the streams around Mute Gill [7560 1930] and by the side of the fell road.

Harthwaite Tuff Formation

The formation is thickest in the southern part of the inlier, near Roman Fell and Keisley, and contains various rock-types including acid tuffs, volcanic sandstones and beds of finely laminated siltstone.

On Roman Fell these rocks crop out in three areas, around Hilton Sike [7490 2527], below High Band [7530 1980] and near Mute Gill [7566 1924]. The oldest beds, seen west of Roman Fell Nook [7492 2038], are volcanic sandstones containing abundant fragments of the underlying Knock Pike Tuff. They are overlain by finely banded siltstones with interbedded tuffs [7487 2033], followed, in the area south of Hilton Sike, by a thick sequence of acid vitric crystal tuffs and volcanic sandstones, the uppermost beds of the Borrowdale sequence preserved hereabouts.

There is a good section [7084 2484 to 7079 2481] through the youngest beds in Harthwaite Sike and scattered exposures on Harthwaite and on Dod Hill, where a small quarry [7103 2524] provides a section in the lowest volcanic sandstones. Farther north, tuffs are poorly exposed in a tributary of Pusgill [7060 2600], and just south of Milburn Beck [6797 2845] where an outcrop of breccia consists mainly of angular fragments of ash-flow tuff in a sandy matrix.

Northern Sequence

At the southern end of the Borrowdale tract east of Melmerby at least 55 m of green basic andesite of Eycott type dip gently eastwards. The markedly amygdaloidal lower part of this section, exposed in the eastern bank of Melmerby Beck [6250 3721], is strongly altered with much secondary carbonate, chlorite and albite, but less altered andesites, with characteristic large pale green labradorite phenocrysts, crop out higher up the bank on Rake Brow [6254 3728].

Northwards these rocks are faulted out, and the beds beyond are not well exposed. Amygdaloidal and vesicular lavas, sporadically exposed in the beck, appear to consist mainly of andesites, interbedded with thin bands of acid lava and lapilli-tuff, but all are strongly altered and difficult to identify, and the bedding is obscure.

At the northern end of the Borrowdale tract dark green basic andesites strike generally east-north-east with high dips. These rocks are best seen in two disused roadstone quarries in Deep Slack [6228 3814 and 6236 3802], particularly in the northern excavation where massive vesicular amygdaloidal andesites dip north-westwards at 45° to 70°. The beds are much faulted, however, and outside the quarries the bedding is again obscure, so that scattered outcrops of lithic and pumice tuff in Shieldgreen Wood [6241 3796, 6228 3813] cannot be related one to another.

Coniston Limestone Group

A post-Borrowdale period of folding, uplift and erosion gave rise to a subdued topography which gradually submerged beneath the sea, and became covered by the shallow-water sediments of the Coniston Limestone Group. In the Cross Fell area the group is divided into the Dufton Shales (Dean 1959) and the overlying Swindale Shales.

The Dufton Shales, about 400 m thick, consist of dark grey, partly calcareous siltstones and mudstones, with thin bands, lenticles or nodules, of silty limestone. Near the base and at the top of the formation, sandier beds occur locally. Thus the lowest part of the Dufton Shales between Milburn Beck and Roman Fell consists of sandstones and sandy siltstones, the '*corona* Beds' facies. These beds contain a variable amount of volcanic detritus, probably derived from nearby outcrops of the Borrowdale Volcanic Group, and may indicate proximity to a contemporaneous shore-line. This facies is not represented near Melmerby (Dean 1959). The highest beds of the Dufton Shales (the '*Diacalymene* Beds') are also sandy, but the detritus is quartzitic rather than volcanic.

The Swindale Shales, about 40 m thick, consist largely of mid-grey calcareous mudstones, with bands and lenticles of greenish grey fine-grained limestone, commonly decalcified at outcrop to brown rottenstone. The proportion of limestone in the Swindale Shales is very variable; at Keisley there is a bioclastic limestone more than 50 m thick, and further north, in Swindale Beck, 19 m of limestone (the Swindale Limestone), including a 5-m mudstone parting near the top, lie at the base of the sequence. In the short section near Melmerby 1.5 m of limestone are present within a mainly mudstone sequence. The Swindale Shales appear everywhere to rest with slight unconformity on the eroded upper surface of the Dufton Shales (Fig. 4).

It would be difficult further to subdivide the lithologically monotonous sequence of the Coniston Limestone Group were it not for its rich and varied shelly faunas which give a detailed biostratigraphy (Figs. 1, 4, 5) spanning the upper part of the Caradoc Series and the whole of the Ashgill Series.

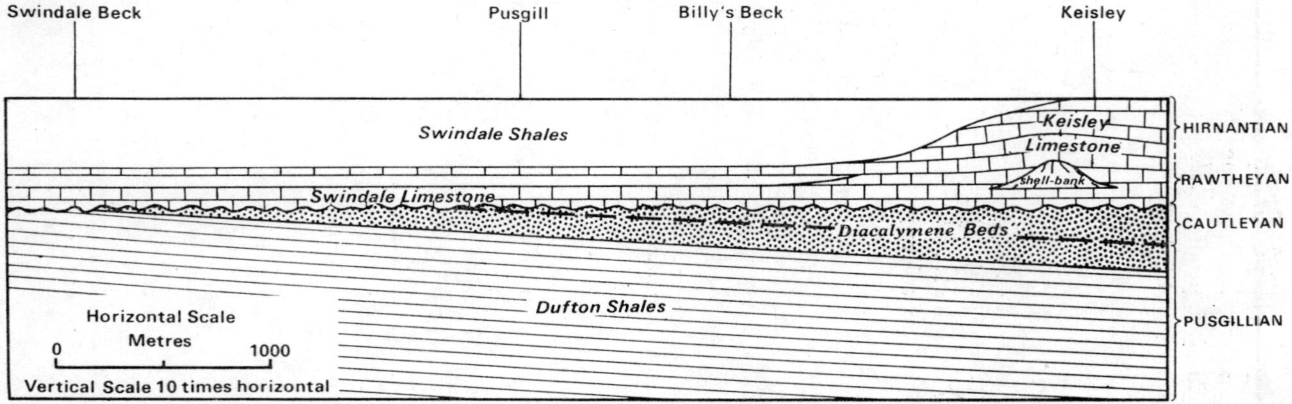

Fig. 4. *The stratigraphical relationship of the Dufton Shales and Swindale Limestone between Swindale Beck (Knock) and Keisley*

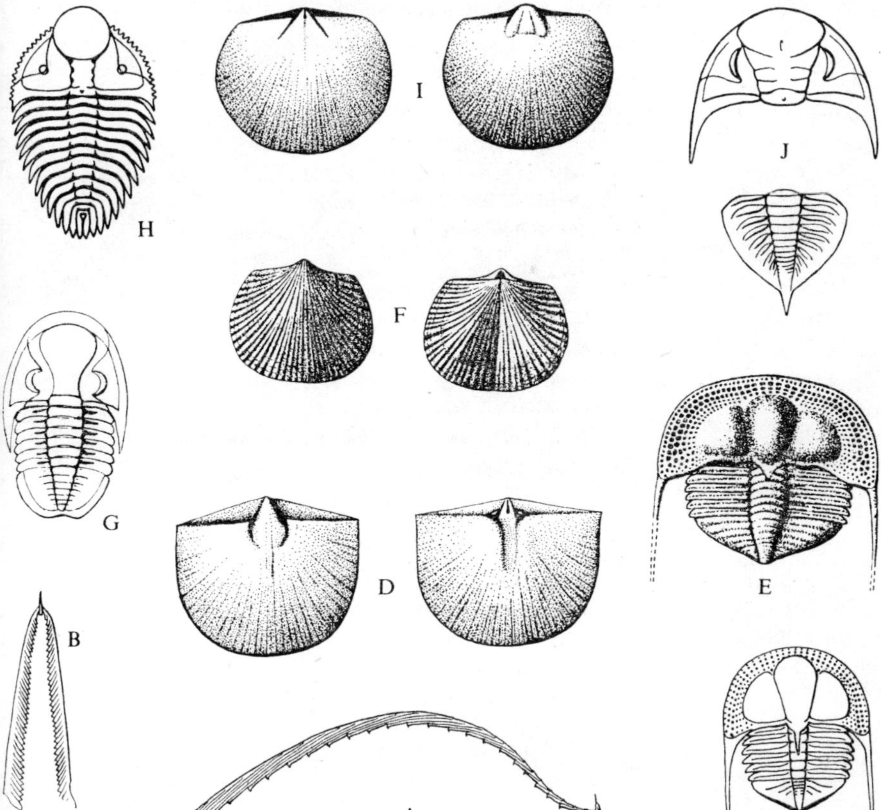

Fig. 5. *Characteristic Ordovician fossils*
(All drawings are natural size except A, which is × 2, and G and H which are × 3. Figs. E and F are reproduced from *British Palaeozoic fossils*, with kind permission of the Trustees of the British Museum (Natural History)).
Skiddaw Group: Llanvirn Series—A, *Nicholsonograptus fasciculatus* (Nicholson), × 2; B, *Didymograptus bifidus* (Hall). Coniston Limestone Group: Caradoc Series: Longvillian Stage—C, *Broeggerolithus nicholsoni* (Reed); D, *Dolerorthis duftonensis* (Reed): Onnian Stage—E, *Onnia gracilis* (Bancroft); F, *Onniella broeggeri* Bancroft: Ashgill Series: Rawtheyan Stage—G, *Phillipsinella parabola* (Barrande), × 3; H, *Staurocephalus clavifrons* Angelin, × 3: Hirnantian Stage—I, *Hirnantia sagittifera* (McCoy); J, *Dalmanitina mucronata* (Brongniart).
(Prepared by Dr. A. W. A. Rushton, Palaeontological Department.)

Typical fossils from each stage are given in the following table, prepared by Dr. A. W. A. Rushton of the Palaeontological Department.

Hirnantian:	*Eostropheodonta hirnantensis*, *Hirnantia sagittifera*, *Dalmanitina mucronata*.
Rawtheyan:	*Christiania sp.*, Plectambonitaceans and Orthids, *Phillipsinella parabola*, *Staurocephalus clavifrons*, *Tretaspis* cf. *radialis*.
minor unconformity	
Cautleyan: (lower part)	Dalmanellids and Orthids, *Sampo ruralis*, *Diacalymene marginata*.
Pusgillian:	*Onniella* cf. *broeggeri*, *Sericoidea sp.*, *Gravicalymene jugifera*, *Onnicalymene onniensis*, *Tretaspis* cf. *moldenensis*.
Onnian:	*Onniella broeggeri*, *Sericoidea sp.*, *Lonchodomus pennatus*, *Onnia spp.*
Actonian:	*Onniella* cf. *aspasia*, *Onnicalymene* cf. *salteri*.
Marshbrookian:	*Broeggerolithus transiens*, *Brongniartella bisulcata*.
Longvillian: (especially upper part)	*Dalmanella spp.*, *Kjaerina spp.*, *Sowerbyella sericea*, *Broeggerolithus nicholsoni*.
Longvillian: ('*corona* Beds' facies)	Horny brachiopods including *Trematis corona*, *Lingulasma tenuigranulum*, bivalves, gastropods and ostracods.

The best sections through the Coniston Limestone Group are briefly described below.

Roman Fell

In the headwaters of Lycum Sike [7496 1996] and nearby streams, reddened '*corona* Beds', vertical or overturned, overlie rhyolitic tuffs. In Hilton Beck, Longvillian mudstones crop out upstream from the Hilton Fault.

Dufton

The best and most accessible section in the lower part of the Dufton Shales is in Harthwaite Sike. Rhyolitic, lithic- and crystal-tuffs forming the highest beds of the Borrowdale Volcanic Group are exposed in the small plantation south of Harthwaite Cottage, dipping steeply west-south-west. Downstream, approaching a fence across the stream

[7078 2480], they give way to tuffaceous sandstones with abundant small calcareous shell fragments and sporadic *Lingulasma*, best seen high on the south bank within the plantation. Beyond the fence, the sandstones are succeeded by grey tuffaceous siltstones with a typical '*corona* Beds' fauna. Farther downstream, beds of Upper Longvillian and Marshbrookian age are exposed.

In Pusgill, '*corona* Beds' rest on tuffs near the Dufton Pike Fault [7054 2600]. Downstream they are faulted against Onnian and Pusgillian mudstones well exposed in the gorge upstream from Pusgill House. The topmost Pusgillian strata, seen west of Hindriggs, are very sandy siltstones. This section, and that in Swindale Beck, are the best exposures of Pusgillian strata in England.

Billy's Beck [7098 2536] also provides a section in high Pusgillian siltstones, and in the overlying '*Diacalymene* Beds' of lower Cautleyan age. The latter are faulted against the Swindale Limestone and Shales.

The Keisley Limestone is well exposed in the disused quarries east of Keisley House. The main part of the limestone is almost devoid of macrofossils other than crinoid debris. A local shell-bank yielding numerous trilobites, especially *Stenopareia*, with abundant brachiopods and other fossils, is exposed north of the wall above the quarries [7140 2390]. In the old (eastern) quarry [7158 2384] the limestones are folded into an anticline, in the core of which are calcareous siltstones possibly of Cautleyan age. In the newer (western) quarry the strata dip consistently to the south. The succeeding beds exposed on a bend in the track leading to this quarry are graptolitic mudstones of early Silurian (*Monograptus atavus* Zone) age (p.28). The limestones appear to be equivalent in age to most of the Swindale Shales farther north.

Knock

The classic section in Knock Gill (Swindale Beck) (Fig. 6), though much faulted, provides exposures of rocks from all the Caradocian and Ashgillian stages except the Onnian, which is faulted out, and the Cautleyan which is cut out by the sub-Swindale Shales unconformity.

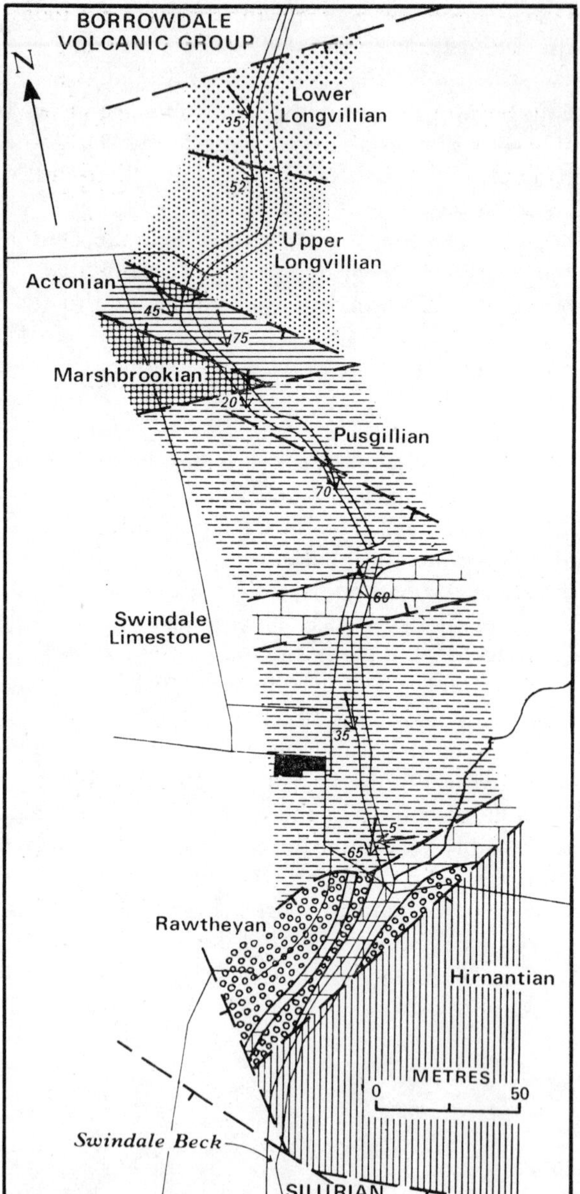

Fig. 6. *Outcrops of the Coniston Limestone Group in Swindale (Beck Knock)*

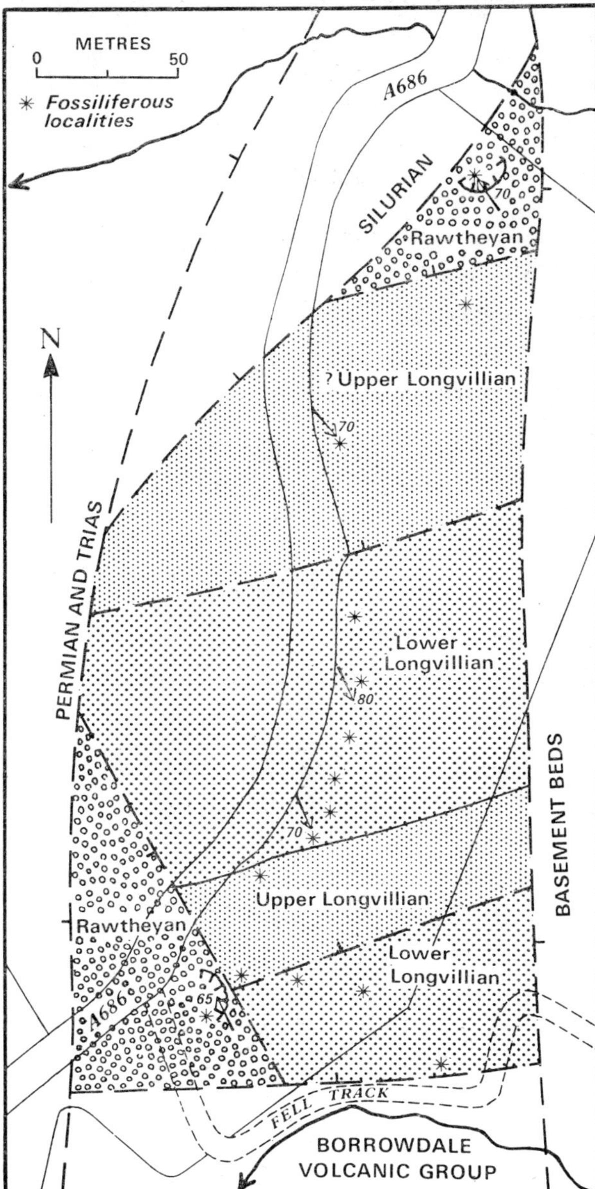

Fig. 7. *Outcrops of the Coniston Limestone Group east of Melmerby*

Melmerby

About 150 m of Dufton Shales, exposed mainly along the roadside (Fig. 7), dip steeply to the south-south-east, and yield both Lower and Upper Longvillian faunas. The Lower Longvillian strata are about 100 m thick. Recent trenching has revealed two small fault-blocks of Swindale Shales with associated thin muddy limestones, in the north-east and south-west corners respectively of the upper Ordovician tract. These beds yield rich Rawtheyan faunas.

3/SILURIAN

Outcrops of Silurian rocks are so restricted and in such heavily faulted ground that they are interpretable only in terms of successions established outside the inlier. Their graptolite faunas indicate that strata of both the Llandovery and Wenlock series are present, although the full sequence of zones established in the Lake District has not been recognized. Fig. 8 gives the standard succession and indicates the range of strata to be seen in the inlier. The Llandovery and basal Wenlock strata contain an abundant and varied graptolite fauna. This is summarized in Figs. 12 and 13, the letters at the column heads referring to localities indicated in Figs. 9–11 (all of which diagrams are reproduced from Burgess and others 1970).

LITHOSTRATIGRAPHIC SUBDIVISION		LOCALITY	ZONE PROVED PRESENT	SERIES	SHELLY FOSSIL STAGE	GRAPTOLITE ZONE
Brathay Flags		K7		WENLOCK		*Cyrtograptus linnarssoni*
		K4—6	?			*Cyrtograptus rigidus* *Monograptus antennularius* *Monograptus riccartonensis* *Cyrtograptus murchisoni*
		S 35—36				*Cyrtograptus centrifugus*
Browgill Beds	Stockdale Shales	S 34 S30—33 S22—29		LLANDOVERY	Telychian	*Monoclimacis crenulata* *Monoclimacis griestoniensis* *Monograptus crispus*
		S7—S21			Fronian	*Monograptus turriculatus*
		S3—S6, R5				[*Rastrites maximus* Subzone]
						Monograptus sedgwickii
Skelgill Shales		R4			Idwian	*Monograptus convolutus*
						Monograptus argenteus [or *Monograptus leptotheca*]
		R3 S1, S2				*Diplograptus magnus* *Monograptus triangulatus*
		R1, R2			Rhuddanian	*Monograptus cyphus* *Monograptus acinaces*
		K1,K2				*Monograptus atavus* *Akidograptus acuminatus* *Glyptograptus persculptus*

Fig. 8. *The Silurian stratigraphy of the Cross Fell Inlier*

As in the Lake District the Llandovery rocks are collectively known as the Stockdale Shales, which are divided on litho-

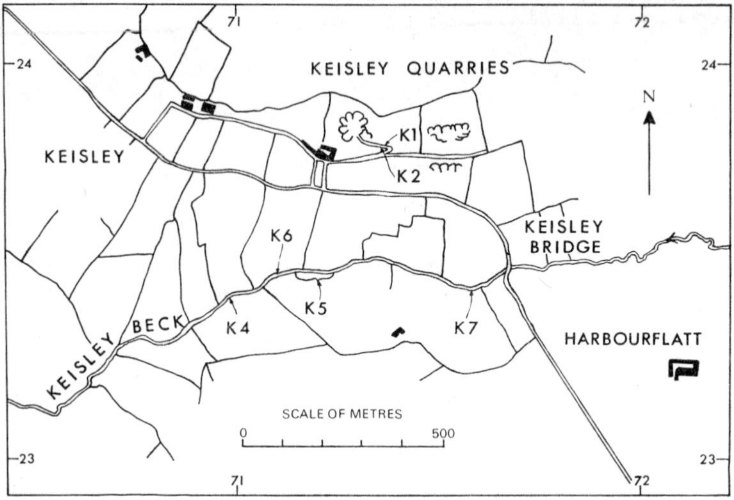

Fig. 9. *Silurian graptolite localities near Keisley*

logical grounds into two formations (Fig. 8). The Skelgill Shales are mainly dark brown to black shaly or blocky graptolitic mudstones. The oldest beds exposed lie at the entrance to Keisley Quarry [7138 2379] where the Keisley Limestone is overlain by shales with limestone nodules, containing a brachiopod fauna probably of Lower Llandovery age (Temple 1968), followed by black mudstones with a graptolite fauna indicative of the *Monograptus atavus* Zone (Fig. 9; K1, K2). Younger beds are seen in isolated exposures in Great Rundale Beck and Swindale Beck, Knock (Fig. 10; R1–4, S1, S2).

The junction with the overlying Browgill Beds is not exposed. The best section in these beds is in Swindale Beck, Knock (Figs. 10, 11). In the lower part of the sequence, comprising the *Monograptus turriculatus*, *M. crispus* and *Monoclimacis griestoniensis* zones, the dominant lithology is pale greenish grey unfossiliferous mudstone. Black graptolitic mudstone is present only as sporadic 1 to 5 cm bands. A few metres stratigraphically above the highest band of the *M. griestoniensis* Zone, the barren grey mudstone is succeeded by red mudstone. (Although the section is here cut by a lamprophyre dyke, the colour change is not related to the intrusion but occurs regionally at this horizon.) The highest Llandovery

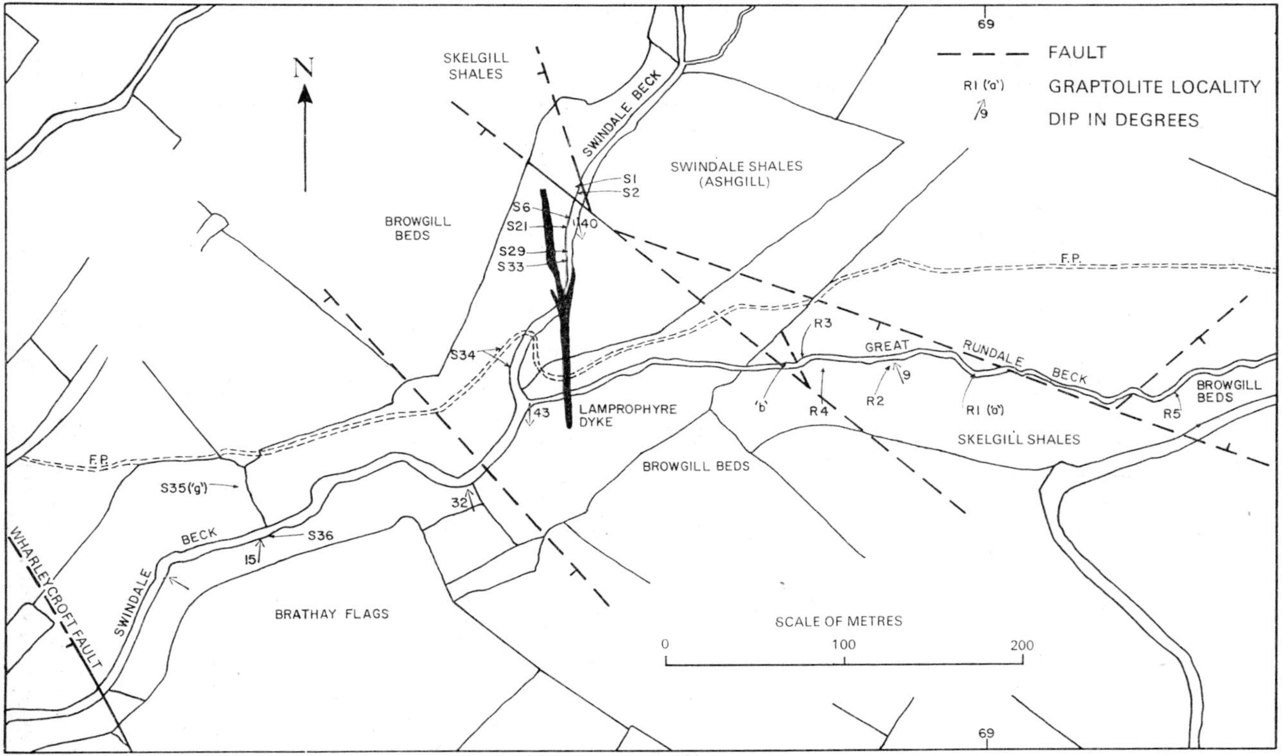

Fig. 10. *Silurian graptolite localities near Swindale Beck (Knock)*

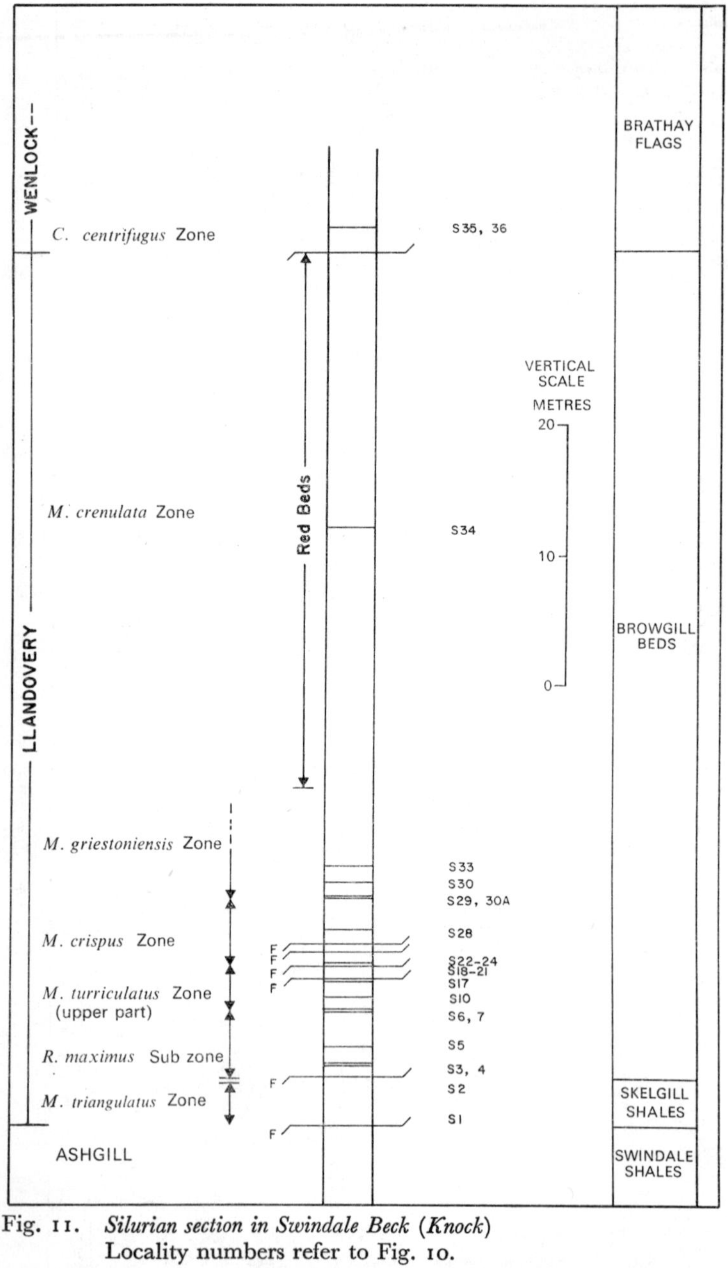

Fig. 11. *Silurian section in Swindale Beck (Knock)*
Locality numbers refer to Fig. 10.

Graptolite Zone Locality No.	*M. atavus* Zone		*M. cyphus* Zone			*M. triangulatus* Zone		*D. magnus* Zone	*M. convolutus* Zone
	K1	K2	R1	R2	'b'	S1	S2	R3	R4
Climacograptus sp.									*
medius	cf								
miserabilis	*	*			*			?	
normalis	*	*						*	eg
rectangularis		*	*		?	*	*	*	*
alternis						*			
t. linearis						*			
Pseudoclimacograptus									
hughesi			*	*	*				
undulatus						*	*	*	*
retroversus								*	*
Diplograptus sp.		*							
modestus		*							
magnus								*	
Cystograptus									
vesiculosus	?	*							
v. penna			*						
Glyptograptus			*						
t. tamariscus			*	sl		*		*	*
sinuatus				*	*		*	eg	
t. angulatus				*			*		
elegans							?		
vas								*	
t. linearis									*
Orthograptus									
mutabilis			*			*			
cyperoides			*			*		*	
bellulus			cf	cf	cf				*
Petalograptus sp.							*		
minor					*			*	
palmeus latus								*	
primulus								*	
Cephalograptus									
cometa									*
Retiolites sp.									*
Raphidograptus sp.									?
extenuatus	*								
toernquisti			*	*	*	*		*	
Monograptus									
atavus		*	*	*	*				
acinaces			?						
cyphus			*	*	*				
gregarius			?		*	*	*	*	*
revolutus			*	*	*	*	*		
sandersoni			*		*				
argutus						*		*	
concinnus						?		*	*
t. separatus						*			
t. triangulatus						*	sl		
c. communis							sl	*	*
leptotheca							*		*
c. rostratus								*	
denticulatus								*	*
t. fimbriatus								*	
convolutus									*
decipiens									*
delicatulus									?
elongatus									eg
limatulus									*
lobiferus									*
Pristiograptus									
regularis									*
Rastrites spp.								*	*
Diversograptus ?									
capillaris									*

Fig. 12. *Lower Llandovery graptolites*
(Identifications by Dr. R. B. Rickards)
Locality numbers refer to Figs. 9 and 10.

Graptolite Zone	M. turriculatus Zone																				M. crispus Zone								M. griestoniensis Zone					M. crenulata Zone	C. centrifugus Zone	
	R. maximus Sub zone					M. turriculatus Zone (upper part).																														
Locality No.	S3	S4	S5	S6	R5	S7	S8	S9	S10	S11	S12	S13	S14	S15	S16	S17	S18	S19	S20	S21	S22	S23	S24	S25	S26	S27	S28	S29	S30A	S30	S31	S32	S33	S34	S35	S36
Climacograptus sp.			*		*															*		?		?												
Glyptograptus									?																											
elegans	*	*	*			?																														
Petalograptus																																				
tenuis			*		*															*					?		*									
Retiolites																																				
g. angustidens																													*	*					*	
g. geinitzianus																																	*	*	*	*
Pseudoplegmatograptus																																				
obesus			*		*																															
elleswoodi																																		*		
Monoclimacis spp.																												*			*	*				
griestoniensis																													*	*	*	*	*			
"cf. griestoniensis"*																													*	*	*		*			
crenulata																																		*		
shottoni																																			*	
v. basilica																																			*	*
v. vomerina																																				*
Monograptus																																				
runcinatus*	*	*	*		*			?				?				*																				
turriculatus	?				*	*		*	*	*	*	*		*																						
exiguus	?	*		*		*			*	*														*	*		*									
balli		*	*	*		*	*	*	*	*	?	*	*	*	*					*																
pseudobecki		*	*							*	*	*	?	*						?																
barrandei*			*																	?																
planus			*	*																						*	*			*	*					
proteus			*			*		*		*			*	*	?		*	*	?	*	?															
spiralis			*																					?	?		?							*		
gemmatus					*							*			?																					
marri														?			?			?		?	*	?	*	*	*	?					?	*		
galaensis																		?		?																
crispus																					*	*	*	*	*		*									
discus																									*	*	*		*		*	*				
priodon																													*	*	*	cf.	cf.	*	*	*
sp. 1																													*							
sp. 2																															*					
t. spiraloides																																	cf			
sp. 3																																		*		
wimani																																		cf.		
danbyi																																			*	*
m. cautleyensis																																			*	
sp. 4																																			*	
Pristiograptus																																				
regularis		*	*	*		*	*	*	*	*	*	*		*	*		*	*	*	*																
nudus																									*	*	*				?		*	*		
denemarkae																																*	*			
Rastrites																																				
linnaei	*																																			
ex. group linnaei		*	*				*		*	*																										
maximus		cf.	cf.	?	*																															
equidistans										s.l.																										
Spirograptus sp.																									*											
Cyrtograptus sp.																																			*	
Diversograptus ?																																				
capillaris			*	*		*													*																	

*Sensu Elles and Wood.

Fig. 13. *Upper Llandovery and Wenlock graptolites*
(Identifications by Dr. R. B. Rickards)
Locality numbers refer to Fig 10.

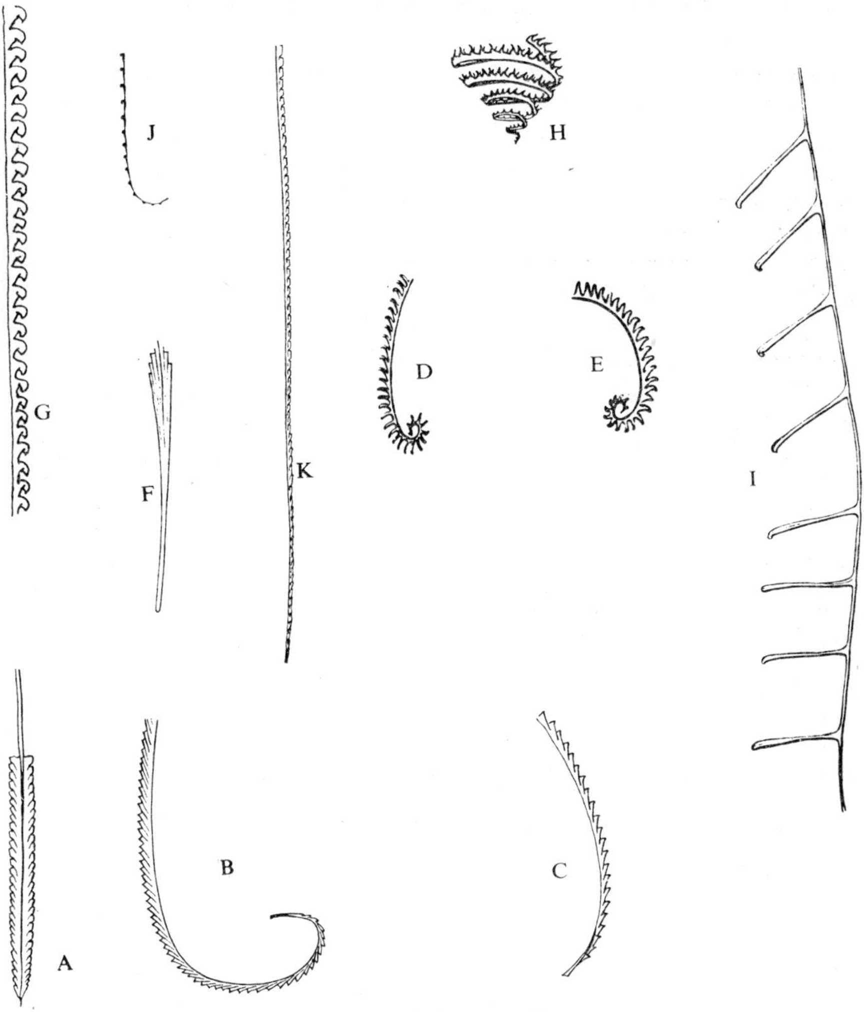

Fig. 14. *Characteristic Silurian graptolites*
(All drawings are natural size).
Skelgill Shales: Llandovery Series (Lower, Middle and Upper)—A, *Diplograptus modestus* Lapworth; B, *Monograptus cyphus* Lapworth; C, *M. gregarius* Lapworth; D, *M. triangulatus triangulatus* (Harkness); E, *M. t. separatus* Sudbury; F, *Cephalograptus cometa* (Geinitz); G, *M. lobiferus* (McCoy). Browgill Beds: Llandovery Series (Upper)—H, *M. turriculatus* (Barrande); I, *Rastrites maximus* Carruthers; J, *M. crispus* Lapworth; K, *Monoclimacis griestoniensis* (Nicol).
(Prepared by Dr. A. W. A. Rushton, Palaeontological Department)

graptolite band lies within the red beds exposed by the path on the west bank of Swindale Beck (Fig. 10; S34). It is indicative of the *Monoclimacis crenulata* Zone, and is the only recorded occurrence of fossils of this age in the north of England.

Browgill Beds are also exposed on the east bank of the A.686 road near Melmerby [6233 3862]. The exposures are in the upper part of the *M. turriculatus* Zone. The section is notable in containing several bands of white clay, probably of volcanic origin, interbedded with the detrital sediments.

The Wenlock strata are lithologically identical with the Brathay Flags of the Lake District–dark bluish grey, laminated, graptolitic siltstones and mudstones with sporadic calcareous nodules. The base of the Flags is not exposed, but strata seen in Swindale Beck (Fig. 10; S35, S36), belonging to the *Cyrtograptus centrifugus* Zone, are low in the sequence.

Higher horizons in the Brathay Flags are exposed downstream from Keisley Bridge (Fig. 9; K4–K6). The fauna includes *Monoclimacis flumendosae*, *Monograptus flemingii* and *Pristiograptus dubius*. In addition, in the vicinity of K7, several exposures have yielded *Cyrtograptus linnarssoni* and *Monograptus flexilis*. These last are indicative of the *C. linnarssoni* Zone; (the former localities may belong to this Zone, but may also be somewhat older). Many of the beds in this section are quite markedly reddened, but this coloration is secondary.

Brathay Flags are also exposed in the fields south of Harbourflatt Farm [7210 2323] and north of Hilton [7361 2123] where they have yielded a fauna indicating the *Monograptus riccartonensis* Zone.

No younger Silurian strata are exposed in the inlier. The beds formerly referred to the Coniston Grits (Ludlow) [6239 3983] (Shotton 1935, p. 646) are now believed to be of Lower Carboniferous age.

Characteristic graptolites from the Silurian rocks are illustrated in Fig. 14.

4/LOWER PALAEOZOIC MINOR INTRUSIONS

Minor intrusions of Lower Palaeozoic age occur throughout the inlier, and are particularly numerous in the northern part. They are mostly steeply-inclined dykes, intruded along joints or cleavages; they remain essentially discordant even where they locally invade the bedding, as this occurs only where dips are steep. Their general trend is Caledonoid (east-north-east), although there are many exceptions. Where the dykes are most numerous, the total crustal extension normal to their trend is estimated at 5 to 10 per cent.

Because thermal spotting comes on eastwards in the northern part of the inlier, it seems possible that the numerous dykes originated from a major granitic intrusion, probably part of the Weardale Granite, underlying the adjacent part of the Alston Block. Some of the isolated microgranitic intrusions farther south, however, may have emanated from the Shap Granite (Hudson 1937).

The Lower Palaeozoic intrusions are grouped into dolerites, lamprophyres and acid porphyries, as described below.

Dolerites

These are dark greenish grey or bluish grey, mottled, massive, fine- to medium-grained rocks, consisting generally of albitized feldspar laths, with chlorite replacing pyroxenes, and much secondary quartz, carbonate and leucoxene. In addition to several minor dykes, they form three principal intrusions marked by extensive crags on Cuns Fell, Catterpallot Hill and Baron Side. These masses are sill-like in form, at least 50 m thick, and dip approximately north-west with the bedding, although their margins are locally discordant. The baked zone of silicified hornfels, 5 to 10 m thick, surrounding them is best seen below the Cuns Fell mass [6483 3662].

The dolerites are the oldest intrusive suite in the inlier. The Cuns Fell mass is cut by later porphyry dykes near the fell summit [6481 3688 and 6482 3692], but the field relations between the dolerites and porphyries are best seen at the western end of Catterpallot Hill [6381 3621], where pink felsitic veins invade the joints of the dark grey dolerite. If, as suggested by Hudson (1937) the soda-rich dolerites are the intrusive equivalents of the spilites in the Kirkland Formation (p. 12), they are of Lower Ordovician age.

Lamprophyres

These are generally grey, purple or brown rocks, with prominent mica flakes, commonly of brown biotite, and small phenocrysts of feldspar or quartz. Where orthoclase can be distinguished from plagioclase as the principal feldspar, the more specific terms minette [e.g. 6328 3646] and kersantite [e.g. 6392 3724, 6959 2890] have been applied respectively, but in most lamprophyres the feldspars are too altered for such subdivision. There is generally so little evidence of baking at the margins of the dykes that a low magma temperature during intrusion is inferred.

Outcrops of lamprophyre dykes generally give rise to prominences in stream banks and to low crags on the fell-side. Such crags north of Dry Sike [6399 3743] have been locally worked for walling-stone.

Although definite evidence is lacking, it seems probable that all the lamprophyres were intruded as a single suite dating from end-Silurian times. The youngest rocks invaded are the Silurian Browgill Beds, cut by a dyke exposed near the junction of Swindale Beck and Great Rundale Beck [6875 2734]. Some of the dykes show a well-developed Caledonoid tectonic foliation, best seen in the intrusion north of Dry Sike [6399 3743], which suggests that they were emplaced before the last phase of compressive end-Silurian earth movements.

The close geographical association of the lamprophyres with the acid porphyries supports the view that these are respectively the basic and acid differentiates of the same magma (Hudson 1937). Indeed, several rocks containing large phenocrysts of feldspar and marginally resorbed quartz, in

Fig. 15. *The intrusions and minor structures in the Skiddaw Group at the northern end of the Cross Fell Inlier*

addition to abundant biotite [6288 3675], appear to be intermediate between lamprophyres and porphyries. Further evidence is seen in Melmerby Beck [6314 3688], where a zoned dyke, about 4 m wide, consists of purplish grey fine-grained lamprophyre at the margins, but towards the centre becomes so pink and feldspathic as to be transitional to a feldspar-porphyry.

Acid Porphyries

The numerous dykes of acid porphyry exposed in the northern part of the inlier are shown in Fig. 15. Subdivisions related to the composition of the principal phenocrysts tend to be grouped geographically; for example, the feldspar-quartz porphyries are concentrated on Thack Moor. It is not clear, however, whether these variations relate to successive intrusive phases of a differentiating magma, or whether they reflect tectonic or geographical factors within a single period of intrusion. The field evidence is consistent with a single phase of intrusion contemporaneous with the emplacement of the lamprophyres at the end of the Silurian.

The porphyry dykes are typically pale grey or pink rocks, usually prominent at outcrop, especially amongst the Skiddaw Slates. They are best exposed north-west of Catterpallot Hill [636 364], on the west side of Lad Slack [651 345] and on Cocklock Scar [655 337]. Farther south, the largest intrusion is the Dufton Microgranite [6933 2683] which cuts the Dufton Shales and is a pink rock with phenocrysts of pink feldspar, rounded quartz and muscovite plates.

5/POLYGENETIC CONGLOMERATE

In the northern part of the district, several local pockets jo coarse clastic sediments of Old Red Sandstone facies underlie the Carboniferous Basement Beds and rest unconformably on an irregular surface eroded in folded Lower Palaeozoic rocks. In these crudely stratified conglomerates, individual boulders are up to 0.5 m across, and the clasts, ranging from subangular to well rounded, are set in a matrix of hematitic sand. A variety of pebbles of local origin is present, together with sporadic boulders of biotite-granite. A skin of hematite —presumably desert varnish—coats the surfaces of most of the clasts; this coating is commonly striated, suggesting that the pebbles have been transported after its formation. The locally steep bedding of the conglomerates is believed to be depositional; and the deposit is interpreted as a series of coalescent alluvial fans derived from higher ground to the west. The age of the Polygenetic Conglomerate is not certain. It has been assigned to the Devonian, but could be of Lower Carboniferous age.

The rocks are well exposed around Melmerby Beck, and particularly in the adjacent fell track [628 370], where about 15 m of conglomerate rest unconformably on highly inclined mudstones of the Skiddaw Group. Other exposures are seen in streams east-north-east of Gamblesby, and in Acorn Sike [646 341] near Ousby Townhead, the locality at which the granite boulders have been found.

6/LOWER CARBONIFEROUS

Lower Carboniferous rocks occupy most of the north-eastern (Alston Block) part of the Cross Fell map, and are seen also on the south-west in several small mainly fault-bounded areas between the Lower Palaeozoic rocks of the inlier and the Permo-Triassic rocks of the Vale of Eden. Lithologically they fall into three main divisions, Basement Beds, Orton Group and Alston Group (Fig. 16).

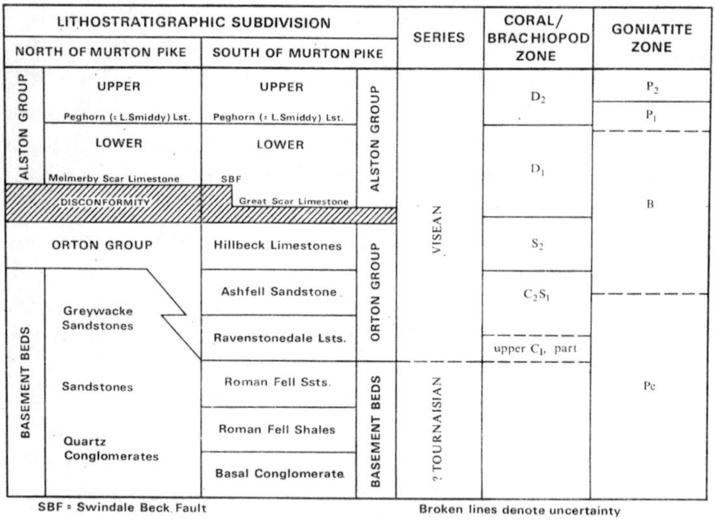

Fig. 16. *The Lower Carboniferous stratigraphy of the Cross Fell Inlier*

The Basement Beds are fluviatile sandstones and quartz conglomerates which contain no zonally diagnostic fossils. They may be partly Viséan in age, but are probably older towards their base, and they rest with marked angular unconformity upon earlier strata. The Orton Group is a sequence of limestones and siltstones with interbedded sandstones. The marine strata yield middle Viséan (S_2) faunas.

The base of the Orton Group, taken at the first appearance of marine strata, is probably diachronous, becoming younger northwards from Murton Pike due to lateral passage into strata of Basement Beds facies. The Alston Group consists of an alternation of limestones, mudstones, siltstones and sandstones, deposited in a rhythmic sequence. At the base is a disconformity, usually marked by erosion and dolomitization of the underlying Orton Group strata; its top is taken at the base of the Great Limestone. It is conveniently split into the lower and upper Alston groups, lying respectively below and above the base of the Peghorn (= Lower Smiddy) Limestone. The group is wholly of D_1 and D_2 age and the base of the upper Alston Group approximates to the base of the D_2 Zone.

The strata are described in detail in several publications, notably those by Garwood (1913), Turner (1927), Shotton (1935), Dunham (1948) and Johnson and Dunham (1963).

Basement Beds

The Basement Beds along the Pennine escarpment are unusually thick compared with adjacent areas in northern England. They are generally thin or absent both to the east on the Alston Block, and to the west around the eastern Lake District, where the only comparable strata lie in the Birk Beck valley, south of Shap. These regional variations may reflect either the presence of a pre-existing valley or a contemporaneous subsidence along the Pennine Fault-zone; as a third possibility, the escarpment sequence may be marginal to a much thicker trough sequence, as yet unproved, beneath the younger rocks of the Vale of Eden. More localized variations also occur. North-west of Murton Pike, considerable thickness variations occur low in the sequence resulting from the infilling of topographic hollows. Also, the succession south of Roman Fell is more than four times as thick as that in High Cup Gill (Fig. 17), the increase affecting all members of the sequence. This increase resulted from contemporaneous movement of a hinge-zone at the northern edge of the Stainmore Trough, close to the Swindale Beck Fault; indeed some movement may have occurred along the fault itself during deposition, but the measurable sections are too far apart to demonstrate this conclusively.

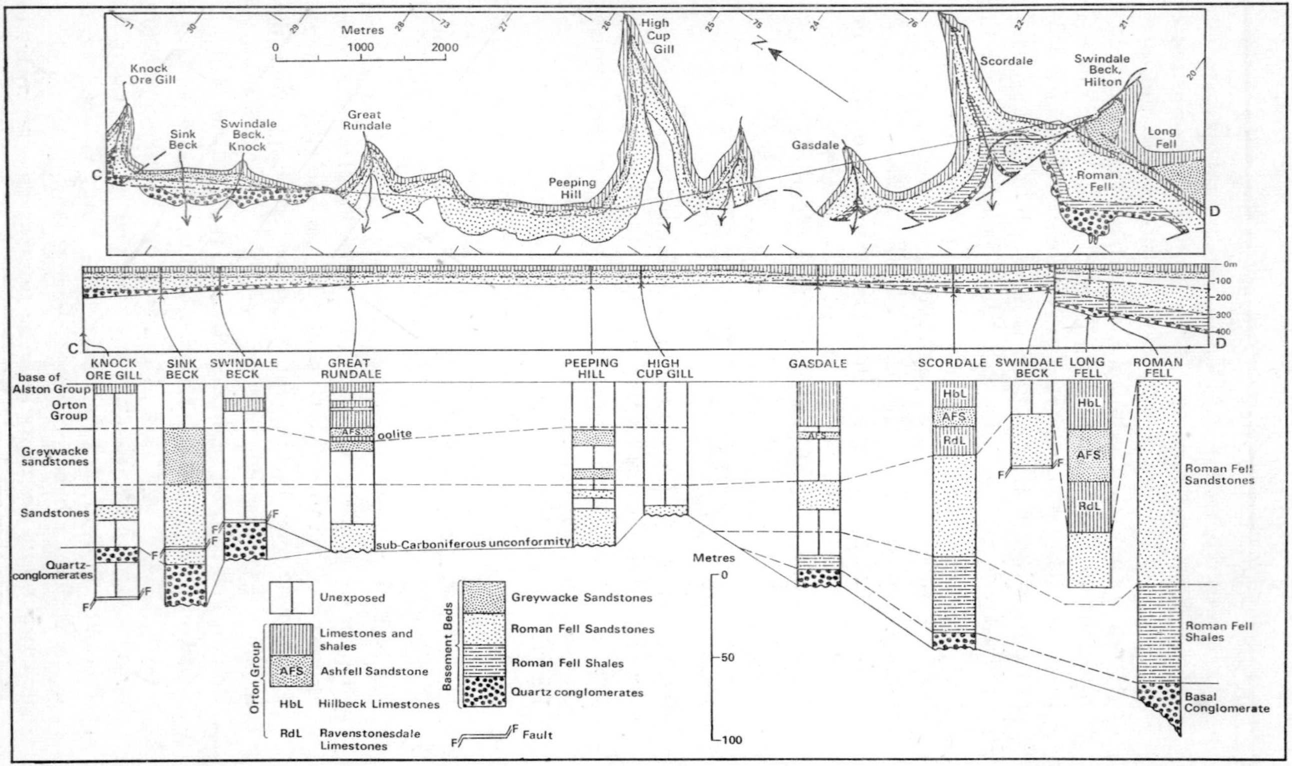

Fig. 17. *The Basement Beds and Orton Group between Roman Fell and Knock Ore Gill*

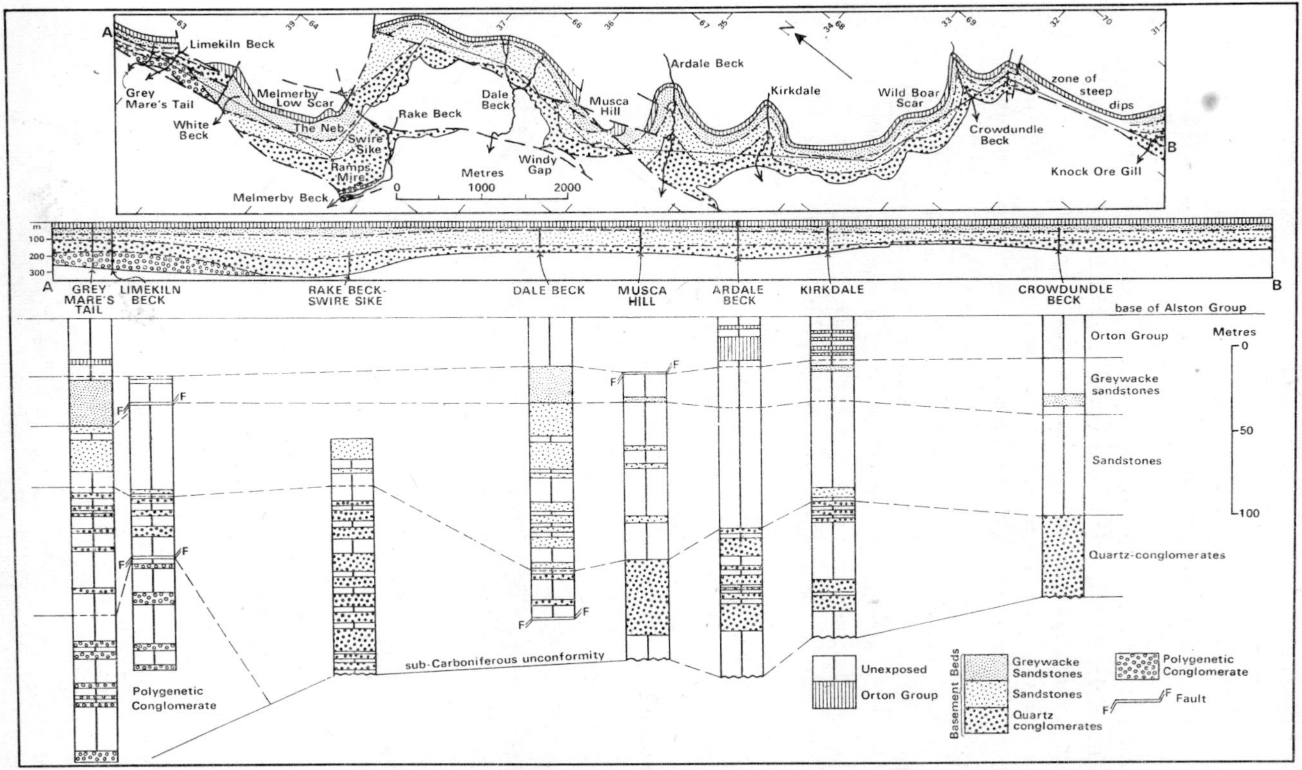

Fig. 18. *The Basement Beds and Orton Group between Knock Ore Gill and Grey Mare's Tail*

The distribution of the various lithofacies is shown diagrammatically in Figs. 17 and 18. Much of the Basement Beds consists of coarse terrigenous clastics, purple-coloured in the north and becoming grey to the south of Crowdundle Beck (Capewell 1956, p. 224). Vein-quartz pebbles are especially abundant, but fragments of mudstone and andesite are also common. In detail, these beds contain local fining-upwards units, generally only a few metres thick and showing a gradational upward passage from conglomerate, through coarse sandstone, into siltstone. The coarser beds are commonly cross-bedded, and rest on erosion surfaces, whilst ripple-marks and plant-debris characterize the finer beds. These features are taken to indicate fluviatile deposition and the sedimentary structures suggest westwards derivation from higher ground near the present Lake District.

In the extreme south, on Roman Fell, the basal conglomerate rests on a surface of considerable topographic relief. The conglomerate consists largely of local clasts and is thickest on the western flank of the fell around Lycum Sike [7500 1990], thinning both northwards and southwards against contemporaneous hills of Lower Palaeozoic tuffs. The maximum thickness here is probably less than 30 m. The overlying Roman Fell Shales consist of purplish red and green siltstones and silty mudstones with thin bands of sandstone and conglomerate. They are poorly exposed generally, the best section lying north of Roman Fell Nook [7536 2060]. The succeeding Roman Fell Sandstones contain numerous quartz pebbles and galls of purple mudstone. The outcrops on the west face of the fell are purplish red in colour, but eastwards rapidly become pink or yellow, indicating that the purplish red coloration, (which also affects the underlying Lower Palaeozoic rocks) is secondary in origin (Burgess and Harrison 1967).

North of the Swindale Beck Fault, the lithological sequence is similar to that in the south, but thinner. At the base, poorly cemented conglomerates rest unconformably upon Skiddaw Group rocks near the confluence of Swindale Beck and Scordale Beck [7510 2138]. Farther west, the unconformity is marked by a line of springs. It is also seen east of the Murton Pike Fault in Murton Beck [7484 2233]. The overlying shales, which thin from 45 m in Swindale to 20 m in Gasdale, are best exposed in Swindale Beck [7550 2140], where 6 m of red and green siltstones and mudstones have yielded scales of the fish '*Rhizodus hibberti*', (Shotton 1935, p. 643). Elsewhere, their outcrop is largely masked by slipped masses of the overlying Roman Fell Sandstones. The best sections of the sandstones, in Swindale Beck [7585 2087] and in Gasdale, show them to thin rapidly northwards.

Between Murton Beck and Great Rundale, strong springs and abundant loose quartz pebbles mark the outcrop of the sub-Carboniferous unconformity whilst the underlying Skiddaw rocks are generally reddened to a depth of several metres. The lower part of the Basement Beds sequence contains interbedded quartz-conglomerates and sandstones, generally more pebbly than the Roman Fell Sandstones to the south, and cropping out in a few short sections on the fell-side. Finer-grained, green sandstones with sporadic conglomerate bands come on higher in the succession, and are well exposed [7194 2506] in the path to High Cup Nick.

North of Brownber, grey quartz-conglomerates form the base of the sequence beneath yellow and grey sandstones. The best section is in Sink Beck [6968 2900], though similar strata are exposed in Knock Ore Gill [6975 3004] and Eller Gill [6898 3125]. Farther north, however, in Crowdundle Beck [6826 3210 to 6835 3230], the conglomerates and overlying sandstones are largely purplish red, the colour change apparently taking place across the Crowdundle Fault. The basal unconformity rises steeply over a buried hill about Littledale and Wythwaite Top where the conglomerates are less than 20 m thick. They rapidly thicken again north of Wythwaite Hole, totalling 80 m in Kirkdale, and are even thicker above the well-exposed unconformity on Cocklock Scar [6544 3388].

A good section in Ardale Beck [6524 3439 to 6558 3454] shows 60 m of Basement Beds including massive conglomerates near the base of the sequence, and fining-upwards units including thin sandstone and siltstone bands farther upstream. To the north, on Windy Gap col, at least 45 m of massive purple conglomerate dip steeply eastwards near the fault and form crags on the northern slopes, whilst the overlying reddish brown sandstones and greywackes are exposed in quarries on Muska Hill. The higher part of the Basement Beds sequence is exposed in the headwaters of Dale Beck [6495 3656 to

6531 3692], where sandstones containing several fining-upwards units are well seen, succeeded by at least 21 m of fine-grained, greyish green, massive, greywacke-sandstones at the top.

The basal unconformity is topographically much lower to the south of Cuns Fell than farther north, due to considerable pre-Carboniferous relief. The base of the sequence is seen in Rake Beck [6330 3706], where 2.7 m of purplish red siltstones with thin bands of coarse sandstone contain rounded mudstone clasts derived from the underlying Skiddaw Group. Although these basal siltstones are absent a short distance downstream [6304 3695], they are again present a little further to the west in the Melmerby fell-track [6287 3704], resting upon the weathered surface of the Polygenetic Conglomerate. Here, the siltstones contain clasts derived from the conglomerate, suggesting an initial reworking of local detritus before the influx of vein-quartz from farther afield. South of the Low Scar, the Basement Beds thicken rapidly westwards to more than 300 m, with a marked pre-Carboniferous slope in this direction.

At the northern end of the inlier, the best sections are through steeply dipping beds in Limekiln Beck and Grey Mare's Tail. Beds seen in Limekiln Beck [6269 3979], were previously assigned to the Silurian Coniston Grits (Shotton 1935, p. 646) but are here correlated with the greywackes high in the Basement Beds. In Grey Mare's Tail [6243 4002 to 6260 4010], about 76 m of purple, well-bedded quartz-conglomerates are overlain by at least 27 m of purplish red and greyish green sandstones, which are in turn succeeded by a similar thickness of massive greywacke-sandstones.

Amongst the much-faulted outcrops of Carboniferous rocks lying west of the Lower Palaeozoic outcrops between Milburn and Melmerby, the Basement Beds are exposed in several sections. These are too short to allow detailed comparison with the escarpment sections. In the sub-Permian oxidation zone, the primary colours are masked by strong purples and yellows.

Orton Group

Most sections in the Orton Group show considerable lithological variation. Thin, dark grey, finely-banded limestones are interbedded with mudstones and sandstones; the clastics are characterized by worm-casts, plant-debris and rootlets, whilst some of the higher limestones are packed with brachiopods and bryozoa. The sequence of lithologies indicates a rapid alternation of shallow-water, marine and non-marine conditions. Throughout most of the district, the uppermost bed of the Orton Group is a dolomitized, nodular limestone, the top of which marks a break in sedimentation and a limited amount of erosion. The disconformity is most marked just north of the Swindale Beck Fault where stronger uplift produced significant erosion between Scordale and Swindale (Fig. 17), but it appears to die out northwards and is not recorded north of Kirkdale.

In the south of the district, the group consists of up to 150 m of marine limestones, sandstones and shales, locally subdivided into Ravenstonedale Limestones at the base, Ashfell Sandstone and Hillbeck Limestones (Burgess and Harrison 1967, p. 209). Up to the top of the Ashfell Sandstone, the sequence covers the upper part of the Lower *Caninia* (C_1) Zone and the Upper *Caninia* (C_2S_1) Zone, whilst the overlying Hillbeck Limestones belong to the *Seminula* (S_2) Zone (Fig. 16).

Near Roman Fell the Ravenstonedale Limestones, consisting of impure limestones, oolites, sandstones and siltstones, crop out in Dobbyhole Gill east of Howgill Fold [7592 1920 to 7600 1925] where a fossiliferous shaly limestone known as the *Thysanophyllum pseudovermiculare* band (Garwood 1913, p. 538; Turner 1927, p. 354) lies about 25 m above the base. North of the Long Fell road, similar impure limestones are cut by a dolerite dyke. Neither the quartzitic Ashfell Sandstone, about 40 m thick, nor the Hillbeck Limestones, apparently about 50 m thick, are well exposed hereabouts.

The group thins to about 80 m just south of the Swindale Beck Fault, the best exposures being in Swindale, where the Ravenstonedale Limestones crop out in the beck [7606 2073] with the Ashfell Sandstone forming Swindale Crag above. North of the fault the group is less than 20 m thick, the higher beds being cut out by the sub-Alston Group unconformity. The size of this break decreases northwards, however, and on the north side of Scordale, all three subdivisions are present again, totalling 45 m. The Ravenstonedale Limestones are seen in Scordale where landslip scars above Lowfield Hush

[7510 2170 and 7526 2182] reveal about 15 m of siltstones with thin limestone bands. The fauna from these beds includes *Michelinia megastoma* and *Syringothyris cuspidata*. The Ashfell Sandstone is less than 10 m thick in Scordale [7526 2182] and is not well exposed farther north, though it forms a small feature in Gasdale. The Hillbeck Limestones are seen at the head of Scordale [7630 2282] and were proved to be almost 30 m thick in the White Mines in Gasdale but elsewhere their surface outcrop is largely obscure. North of Murton Pike, the Ravenstonedale Limestones die out, apparently by lateral passage into sandstones forming part of the Basement Beds. On the south side of Great Rundale [7114 2709] the Ashfell Sandstone is only 6 m thick and rests on a bed of coarse quartz conglomerate. Within the conglomerate is 0.3 m of oolitic limestone containing *Syringothyris cuspidata*. This conglomerate may be the local representative of the Brownber Pebble Bed, a conspicuous marker within the Ravenstonedale Limestones in the southern part of the Vale of Eden. On the north side of Great Rundale [712 275], about 28 m of thin limestones, siltstones and sandstones, yielding S_2 Zone faunas are well exposed; the uppermost beds are dolomitized. Apart from short sections in Swindale Beck and Knock Ore Gill, the group is not well seen again south of Wildboar Scar [6820 3238 to 6821 3236], where the top 5 m form steep crags, and the uppermost limestone is again dolomitic.

Farther north, the group is seen in Kirkdale [6677 3402 to 6669 3398] and in Ardale, which provides the best sections hereabouts. The lower part of the sequence exposed in Ardale Beck [6612 3494 to 6616 3496] consists of dark limestones interbedded with sandstones and seatearths, whilst the upper part, best exposed in Ranscleugh Sike [6613 3484 to 6610 3485], contains thin limestone bands crowded with brachiopods and bryozoa.

To the north-west, the group is largely unexposed, save for short sections below Melmerby Low Scar [6281 3877], in steeply dipping beds in Limekiln Beck [6223 3977], and at the head of Grey Mare's Tail [6260 4010], but an overall thickness of at least 30 m is maintained.

In the Carboniferous tract west of the inlier, there are no complete sections, but scattered outcrops suggest that the general sequence resembles that of the Cross Fell escarpment.

Alston Group

The sequence of strata in the Alston Group is shown in Figs. 19 and 20 which illustrate the best-exposed sections in the district. In the lower Alston Group limestone is the commonest lithology accompanied by subordinate sandstone, siltstone and mudstone partings. In the upper Alston Group, however, terrigenous clastics are much commoner and limestones usually make up less than 25 per cent of the succession. The limestones are in places highly fossiliferous, and some of the commoner species are illustrated in Fig. 21. The units of the rhythmic sequence—cyclothems (Dunham 1948, p. 14)—comprise the following lithologies in upward order: limestone, mudstone, siltstone, flaggy sandstone, massive sandstone, seat-earth, coal. This pattern, with minor variations, is repeated many times on varying scales of thickness. The thickest named limestones (Fig. 20) mark the major cyclothems which are, in the main, laterally the most extensive, though each may contain several minor cyclothems with thinner limestone members. The limestone names are mostly those used by the lead miners of Alston Moor (Forster 1809).

The rocks of the Alston Group, gently inclined to the east, are best seen on the Pennine escarpment, where the massive limestones low in the group form conspicuous scars which contrast with the more subdued and generally peat-covered ground farther east. They are also seen at the foot of the scarp, where they rise from beneath the Permian strata of the Vale of Eden, but exposures are poor here due to the thick cover of glacial drift, and the rocks are patchily reddened, dolomitized and cut by many faults.

North of the Swindale Beck Fault the group is about 300 m thick around Little Fell, and farther north it thickens gradually, reaching about 350 m to the north-west of Cross Fell. South of the Swindale Beck Fault a marked increase in thickness of the beds in the lowest part of the sequence coincides with the line of the fault. In the down-faulted region west of the inlier, sections are few, but thicknesses appear comparable to those on the adjacent parts of the block.

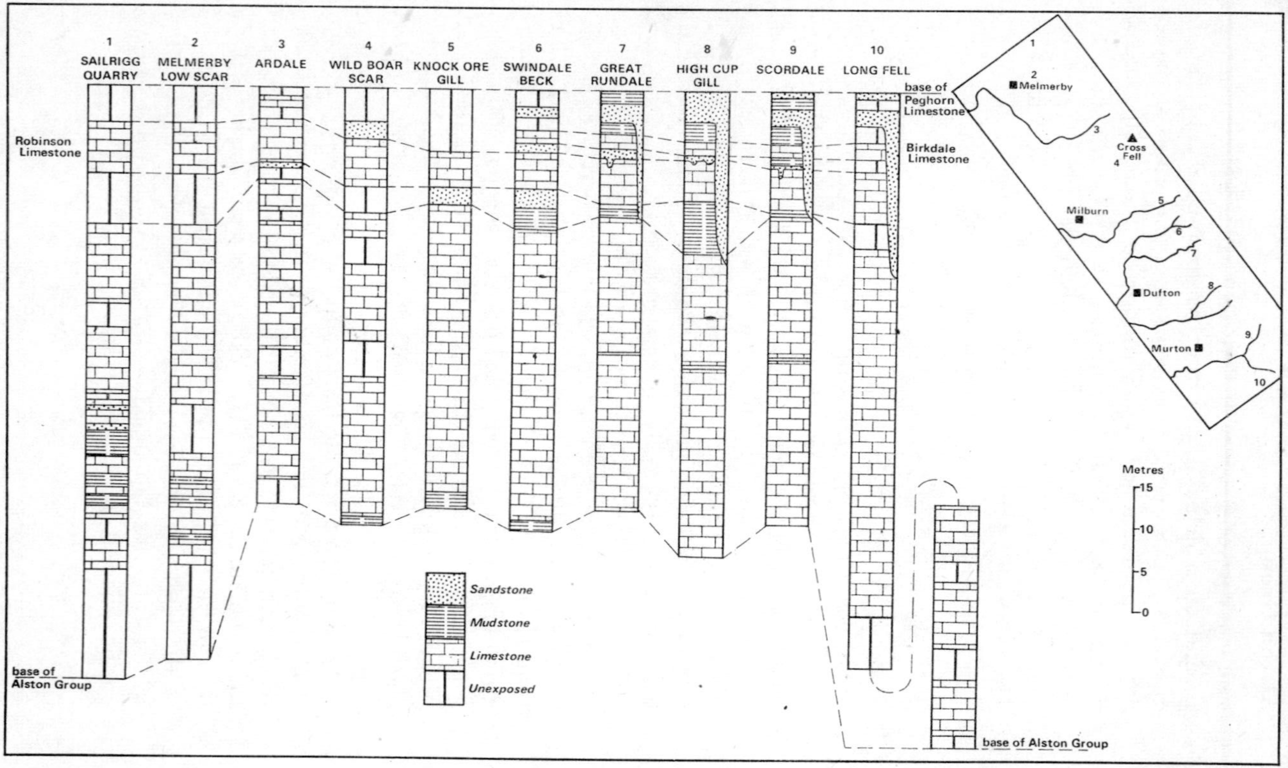

Fig. 19. *Comparative sections in the lower Alston Group*

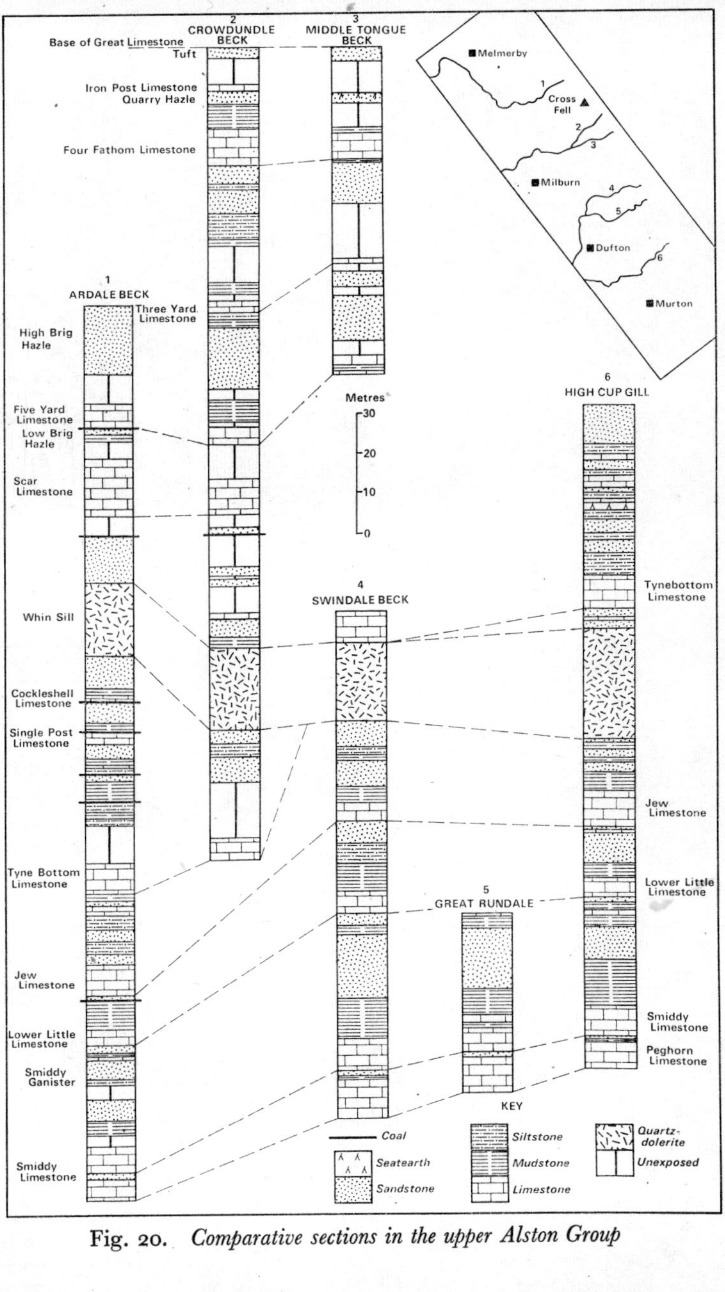

Fig. 20. *Comparative sections in the upper Alston Group*

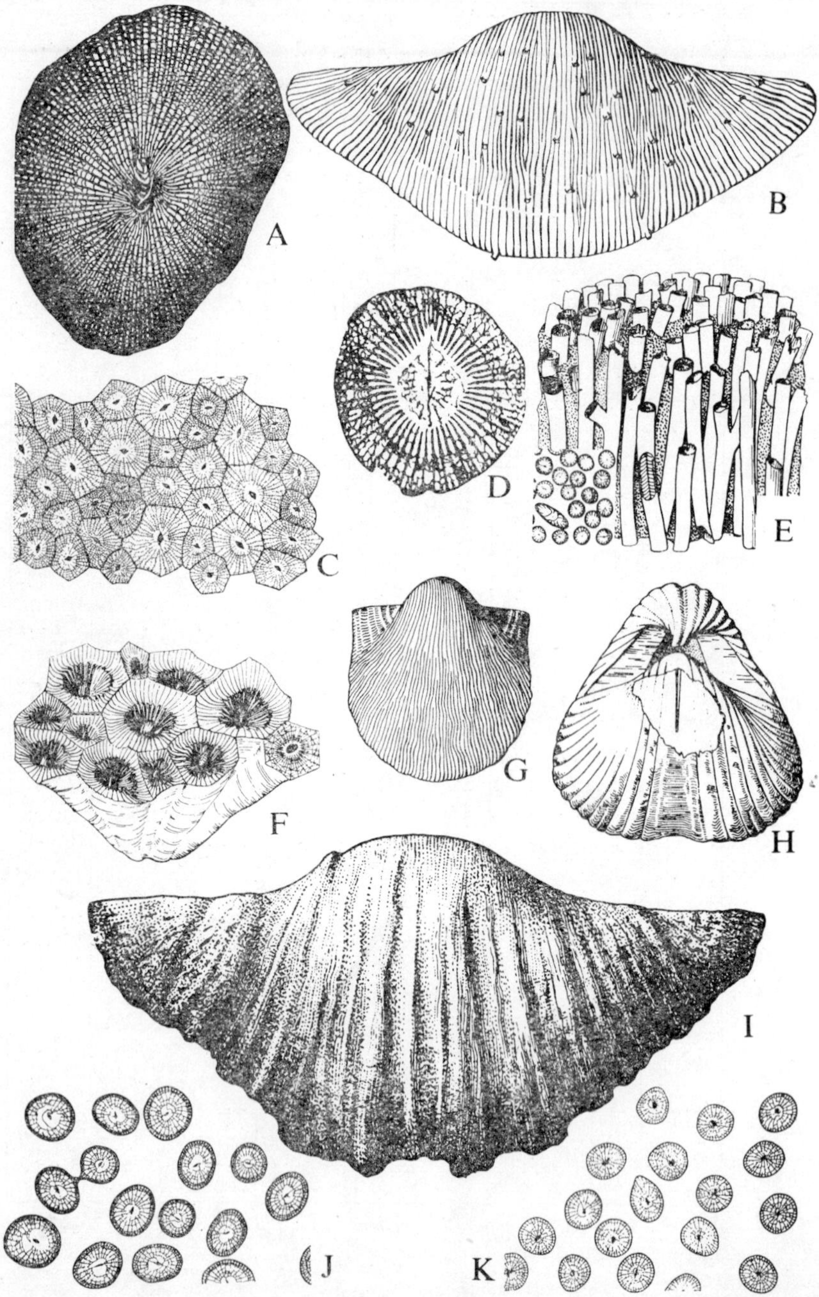

Fig. 21. *Characteristic Lower Carboniferous fossils*

Lower Alston Group

South of the Swindale Beck Fault, the highest Orton Group strata are followed by 18 m of alternating dark grey limestones and siltstones, exposed on the north face of Long Fell. Above these beds rise the massive scars of the pale grey or buff, partly pseudobrecciated Great Scar Limestone, here over 60 m thick.

North of the Swindale Beck Fault, the equivalent beds are much thinner. Along most of the escarpment the lowest formation, the Melmerby Scar Limestone, consists of 35 to 38 m of pale grey to buff, bioclastic limestone, apparently corresponding to the upper part of the Great Scar Limestone to the south. Beneath the limestone, some sections show a thin bed of siltstone which commonly contains limestone nodules and which rests on the dolomitic upper part of the Orton Group. This siltstone, possibly the condensed equivalent of all the lower part of the Great Scar Limestone south of the Swindale Beck Fault, is exposed at the head of Scordale [7632 2280], in Great Rundale [7124 2752], in Knock Ore Gill [7023 3009] and below Wild Boar Scar [6817 3244].

In most sections along the escarpment, major bedding planes with some mudstone lie at about 20 and 33 m above the base of the Melmerby Scar Limestone and commonly form marked steps in the limestone feature. In the north, the lower part of the Melmerby Scar Limestone appears to change in facies across the line of the High Scar Fault. Thus, on Melmerby Low Scar [6293 3853] dark limestones at this horizon are split by several thin mudstone and sandstone partings, which thicken northwards (Fig. 19).

The Melmerby Scar Limestone is commonly very fossiliferous; corals recorded include caninioids, clisiophylloids,

---

Explanation to Fig. 21

(All drawings are natural size except I which is × ½. Figs. A and I are reproduced from *British Palaeozoic fossils*, with kind permission of the Trustees of the British Museum (Natural History)).

Viséan Series—A, *Palaeosmilia murchisoni* Milne Edwards and Haime; B, *Semiplanus latissimus* (J. Sowerby); C, *Lithostrotion portlocki* (Bronn); D, *Dibunophyllum bipartitum* (McCoy); E, *Lithostrotion junceum* (Fleming); F, *Lonsdaleia floriformis* (Martin); G, *Linoprotonia corrugatohemispherica* (Vaughan); H, *Davidsonina carbonaria* (McCoy); I, *Gigantoproductus giganteus* (J. Sowerby); J, *Lithostrotion martini* Milne Edwards and Haime; K, *Lithostrotion pauciradiale* (McCoy). (Prepared by Mr. M. Mitchell, Palaeontological Department.)

Carcinophyllum vaughani, *Dibunophyllum bourtonense*, *Lithostrotion junceum*, *L. maccoyanum*, *L. martini*, *L. portlocki* and *Palaeosmilia murchisoni*. *Lithostrotion* colonies are particularly abundant in a quarry east of the Alston road [6281 3979]. Well-preserved brachiopods are less common, except in a bed about 8 to 10 m below the top of the limestone, the '*Davidsonina septosa* band' which contains in addition to the name fossil, *Delepinea comoides*, *Echinoconchus sp.*, *Gigantoproductus sp.*, *Linoprotonia sp.*, *Megachonetes sp.* and *Punctospirifer sp.* (Johnson and Dunham 1963, p. 28).

Above the Melmerby Scar Limestone the terrigenous beds are variable in thickness and lithology (Fig. 19). The succeeding Robinson Limestone, up to 6 m thick, is pale grey and massive, like the Melmerby Scar Limestone. It contains a scattered fauna of corals and commonly considerable numbers of large shells of *Linoprotonia hemisphaerica*.

The succeeding clastic beds are thin, and consist mainly of sandstone, which in several localities (Fig. 19) fills potholes in the eroded upper surface of the Robinson Limestone. The Birkdale Limestone, a thin dark grey bioclastic limestone with a reddened top, persists throughout the area. It contains few fossils, apart from *Lithostrotion junceum*, and is succeeded by up to 7 m of siltstones and thin sandstones. At several localities in the southern part of the district, much thicker sandstones fill deep channels eroded below this horizon. The channels, probably part of a single large deltaic distributary, are in places more than 100 m wide and up to 20 m deep, cutting down through the Birkdale and Robinson limestones, into the Melmerby Scar Limestone. The best exposures are on the north side of High Cup Gill [7275 2506] and in Scordale [7575 2250].

Upper Alston Group

The clastic sequence separating the Peghorn and Smiddy limestones is too thin to be shown on the 1:25 000 map, and the limestones are thus mapped jointly as the Smiddy Limestone. However, the Peghorn Limestone can be distinguished in good sections (Fig. 20). For example, in Swindale Beck (Knock) [7072 2876], the limestone, 9.5 m thick, lies about 2 m below the Smiddy. Much of the Peghorn Limestone here consists of dark grey, compact bioclastic limestone, characterized by *Lonsdaleia floriformis floriformis*, *Lithostrotion pauciradiale* and transverse gigantoproductoids, but with a 2.5 m band of pale grey partly pseudobrecciated limestone lying about 5 m above the base. Algal nodules of '*Girvanella*' encrusting shell fragments, are common in a band at the top of the limestone. Both the lithological sequence and the '*Girvanella*' band can be traced throughout the district.

The succeeding Smiddy Limestone consists of 8 m of dark grey limestone, and lies at the base of the first thick cyclothem. '*Girvanella*' nodules are common in this bed also, usually in two bands lying 1.5 and 6.5 m above the base. The succeeding mudstones are fossiliferous in the lower part, and grade upwards through sandy siltstones into a thick massive sandstone (the Smiddy Ganister), which forms a strong feature. The sandstone is commonly cambered downslope. The overlying beds include a persistent minor cyclothem, marked at the base by about 1 m of limestone, best seen in Scordale, High Cup Gill and Ardale. In the Low Middle Tongue Level [6904 3210] of Silverband Mine the limestone is underlain by a coal, 0.3 m thick, unusual at this horizon.

The Lower Little Limestone, generally about 4 m thick, is well exposed in Scordale, High Cup Gill, Rundale, Swindale Beck (Knock) and Ardale. It is dark grey, bioclastic, and commonly contains gigantoproductoids and compound corals; '*Girvanella*' nodules are present in some sections, as in Ardale Beck [6644 3522]. The overlying shales and sandstones are very variable and may include a minor cyclothem marked by a thin sandy limestone.

The Jew Limestone consists of about 8 m of dark grey bioclastic limestone commonly sandy towards the base, with many solitary and colonial corals and brachiopods, including *Aulophyllum fungites pachyendothecum*, *Dibunophyllum bipartitum*, *Lithostrotion junceum*, *L. pauciradiale* and gigantoproductoids. It is overlain by alternating shales and sandstones, and a limestone 1 to 2 m thick is present towards the top of the sequence. The latter is best seen near High Cup Nick [747 264] and in Ardale [665 352].

The Tynebottom Limestone, about 9 m thick, closely resembles the Jew in lithology and fauna, but is characterized by a thin band of *Saccamminopsis* lying about 2 m below the top.

Although the microfossil is present in many of the limestones of the upper Alston Group, this band is sufficiently persistent to be used for local correlation. The limestone is best seen on the wide outcrop on High Cup Plain [745 262].

Between the Tynebottom and Scar Limestones the beds are very variable and contain up to seven thin cyclothems, each including a bed of limestone or marine shale; they are known accordingly as the Alternating Beds. They are best seen at the head of Ardale [6657 3544 to 6648 3540], where six separate cyclothems can be distinguished. Discontinuous sections through these beds are also present in Scordale, Maize Beck and in small streams just north of High Cup Plain. Only three limestones from this sequence, the Maize Beck, Single Post and Cockleshell limestones, can be traced for any distance, though others can be identified in good sections. The mudstones immediately above the Tynebottom Limestone contain many large ironstone nodules, which were formerly worked just south of the Dun Fell road [707 298]. The succeeding siltstones and sandstones are poorly exposed. The Maize Beck Limestone consists of 1 m of dark grey bioclastic limestone, usually with a poor fauna including *L. junceum* and small brachiopods. The Single Post Limestone generally crops out as a single bed, about 3 m thick, of pale grey, pseudobrecciated limestone with an irregular top, which weathers to rubble. In contrast, the Cockleshell Limestone consists of dark grey bioclastic limestone, characterized by many large gigantoproductoids and abundant *L. junceum*. A coal lying a short distance below the Scar Limestone has been worked at intervals between Swindale Beck (Knock) and Ardale Beck.

West of the inlier, the section exposed in Ardale Beck [6828 3445 to 6424 3438] probably includes the Alternating Beds sequence (Shotton 1935). The Single Post Limestone is present, about 3 m thick, and a locally worked coal lies close below the Scar Limestone.

The Scar Limestone is composed of mid-grey, wavy-bedded bioclastic limestone, with a rich coral-brachiopod fauna, including a characteristic band of *L. junceum* near the base. The limestone thickens northwards from 5 to nearly 15 m, although on Melmerby Fell it is locally absent. The overlying massive sandstone (the Low Brig Hazle) is well seen at the head of Ardale [6661 3546] although it is thicker farther south where it forms extensive cambered scars.

The succeeding Five Yard Limestone, up to 5 m thick, is generally of darker grey limestone than the Scar, but has a similar fauna. The best sections are seen in Middle Tongue, Crowdundle and Ardale becks. The overlying massive, medium-grained sandstone (the High Brig Hazle), caps much of the Pennine escarpment. Its outcrop is extensively cambered.

The Three Yard Limestone consists of about 3 m of dark grey, often sandy, bioclastic limestone, but is rarely seen at outcrop, being more usually indicated by a row of swallow holes. The best section is in Crowdundle Beck [6942 3318].

The mid-grey, fine-grained, evenly bedded Four Fathom Limestone is best seen in Crowdundle Beck and Middle Tongue Beck. Scattered chert nodules occur throughout the 9 m limestone, which also contains solitary corals, including *Aulophyllum fungites pachyendothecum*, *Caninia sp.*, *Dibunophyllum bipartitum* and brachiopods.

Crowdundle Beck also provides the best section through the succeeding beds up to the Great Limestone [695 333]. Here about 7 m of highly fossiliferous calcareous mudstone overlie the limestone and are succeeded by a thin sandstone (the Quarry Hazle) capped by a seatearth. The Iron Post Limestone, 0.5 m of dark grey, muddy, crinoidal limestone, is overlain in turn by fossiliferous mudstones, a thin sandstone (the Tuft) and a seatearth lying at the base of the Great Limestone. The Iron Post Limestone and overlying shales are also seen in several small streams at the south end of Little Fell.

7/UPPER CARBONIFEROUS

In the Cross Fell area, the only Upper Carboniferous rocks preserved belong to the lowest part of the Millstone Grit or Namurian Series. The limits both of the Namurian and of its constituent stages are defined by reference to particular goniatite bands that occur widely over much of Western Europe. In the Northern Pennines, however, goniatites are extremely rare, and consequently correlation is indirect and somewhat speculative. Following Johnson and others (1962) the base of the Series is currently taken at the base of the Great Limestone and the subdivision between the Pendleian (E_1) and Arnsbergian (E_2) stages at the base of the Lower Fell Top Limestone.

Namurian rocks are restricted at outcrop to a fell-top outlier, lying generally above 2400 ft (732 m) OD, and including Cross Fell and the Dun Fells, to a few small fault blocks near Ousby and Roman Fell, and a small outlier capping Little Fell. The preserved sequence is at least 170 m thick and continues the rhythmic sedimentary pattern of the higher Viséan rocks beneath (Fig. 22) (Johnson and Dunham 1963).

Cross Fell

The harder beds between the Great Limestone and the Firestone Sill are exposed in Middle Tongue Beck [7027 3239 to 7060 3260] and in Crowdundle Beck [6955 3337 to 6999 3381]. Sections which reveal more of the softer mudstones in the sequence are partly artificial. The best, apparently continuous between the Little Limestone and the Coalcleugh Coal, is in the northern wall of Dun Fell Hush [7149 3186 to 7120 3182], a spectacular gully cut by the repeated flushing of water down the outcrop of a mineral vein by early lead miners. This

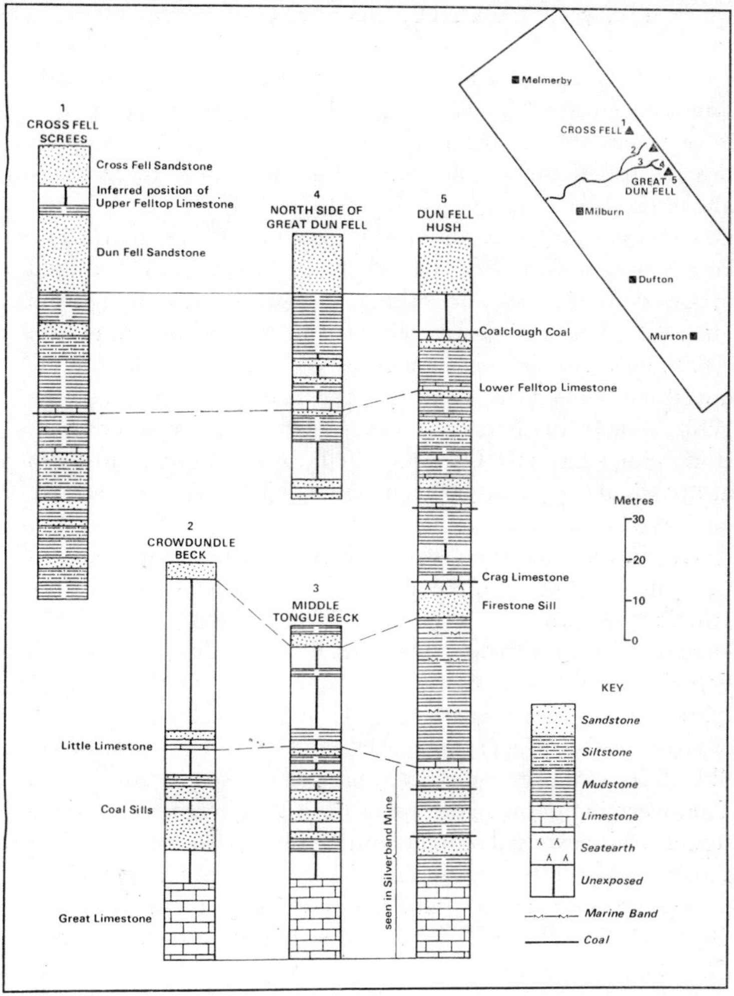

Fig. 22. *Comparative sections in the Millstone Grit*

hushing technique was carried out on a much smaller scale during the resurvey in several small gullies on the shale slopes of The Screes [6883 3501 to 6935 3485] and produced temporary sections in which a continuous sequence was recorded. A small stream was similarly cleared to provide a section on the northern slope of Great Dun Fell.

The Great Limestone forms a prominent feature across the western slopes of Cross Fell, except near the Ardale Head Vein where it is largely replaced by limonite. The best exposures are in Crowdundle Beck [6955 3337] and Middle Tongue Beck [7027 3240] where 18 to 19 m of grey, fine-grained, bioclastic limestone crop out. The Frosterley Band (Dunham 1948, p. 27), is represented by two thin beds of coral- and brachiopod-bearing limestone, locally lying 11.3 and 12.2 m respectively above the base of the Great Limestone. It is seen both in the above stream-sections and on Green Fell [6674 3610]. The largely arenaceous upper part of the cyclothem is well exposed in Crowdundle Beck and Middle Tongue Beck, although the most complete section was recorded in Silverband Mine (op. cit., p. 29). The sequence generally contains three identifiable sandstones—the Low Coal Sill, the High Coal Sill and the White Hazle in ascending order—separated by mudstones containing thin coals. On Green Fell, the two lower sandstones are particularly thick and the Low Coal Sill forms extensive cambered outcrops.

The Little Limestone is seen in continuous section only in Crowdundle Beck, where it comprises 0.8 m of dark grey crinoidal limestone. It is slightly thicker—about 1.2 m—in the wide outcrops north-west of Cross Fell. Southwards, in Dun Fell Hush, it is largely replaced by iron oxides. Thin sandstones lying just above the limestone in Crowdundle Beck and south of Green Fell [677 357] appear to correlate with the Pattinson Sill of the Brampton district, and die out to the south-east. The overlying 33-m sequence of dark grey ferruginous mudstones is fully exposed only in Dun Fell Hush, where it contains three bands rich in marine shells. The highest of these yielded *Cravenoceras sp.* (Johnson and Dunham 1963, p. 54) on Little Dun Fell just beyond the eastern margin of the district. Elsewhere exposures are rare. Beds high in this cyclothem crop out in a stream west of The Screes [6870 3513], whilst 9 m of lower beds are exposed just north of Crowdundle Beck [6964 3356 to 6965 3361], but neither section has yielded marine fossils. Near the top of the cyclothem the Firestone Sill consists of 6 m of medium-grained, cross-bedded sandstone in the stream west of The Screes and forms an extensive rock-pavement a short distance to the west [680 354]. Farther south, it is 4 to 6 m thick and well exposed in Crowdundle Beck, Middle Tongue Beck and Dun Fell Hush. The cyclothem ends with the thin Crag Coal.

Outcrops of the Crag Limestone are restricted to Dun Fell Hush, where 0.36 m of fossiliferous sandy limestone overlies the Crag Coal. Loose blocks of decalcified limestone lying just above the Firestone Sill in Middle Tongue Beck may mark the outcrop. The overlying sequence up to the Lower Fell Top Limestone is 45 m thick on the well exposed Screes section and is mainly composed of mudstone, but it contains two thin sandstones, locally mappable around Cross Fell, lying 12 and 36 m respectively above the Crag Limestone. Just above and below the upper sandstone, the mudstones are rich in marine fossils. The sequence in Dun Fell Hush is similar, although with fewer sandy beds.

The Lower Fell Top Limestone is exposed on The Screes as 0.9 m of medium to dark grey limestone [6934 3490] and in Dun Fell Hush as 0.3 m of brown fine-grained limestone. Along much of its outcrop, however, it is decalcified to a soft yellow 'famp'. The limestone is overlain in The Screes section by about 26 m of poorly exposed fossiliferous mudstones. The sequence above the limestone is better seen at the head of Dun Fell Hush [7129 3184 to 7120 3182] (Fig. 22). About 9 m above the top of this section lies the base of the Dun Fell Sandstone, a medium-grained massive bed. The same sandstone, 18 m thick, forms the strong shelf-like feature just below the summit of Cross Fell.

The Upper Fell Top Limestone is not exposed, but probably closely overlies the Dun Fell Sandstone. It may therefore be present on Cross Fell, within a 6 m interval separating the top of the sandstone from a 9 m bed of flaggy sandstone capping the summit and here termed the Cross Fell Sandstone. Shallow pits near the summits of Cross Fell [6874 3455] and Dun Fell [7107 3196] have been interpreted as pot-holes in the limestone (Johnson and Dunham 1963, p. 58), but excavations on both sites failed to reveal limestone, and it seems more likely that these pits are sandstone diggings.

Ousby

The much faulted and largely drift-covered Namurian outcrops east of Ousby are difficult to correlate with the Pennine sequence. The most informative section is in the eastern part of Star Hows quarry (Shotton 1935, p. 664), [6430 3447], and is as follows:

	m
Mudstone, dark grey, reddened silty; nodular ironstone bands; *Cravenoceras sp.*	4.27
Limestone, dark grey, fine-grained	0.15
Mudstone, dark grey, calcareous	0.15
Limestone, dark grey, fine-grained	0.53
Mudstone, dark grey, calcareous	0.15

The specimens of *Cravenoceras sp.* show sutures and ornamentation comparable with the *C. malhamense/cowlingense* group but, because their inner whorls are not preserved, they cannot be determined with certainty. Although the presence of *Cravenoceras* of this group indicates an E_1 or low E_2 age, the precise stratigraphical horizon of the section is thus not known, though it may correlate with the mudstones beneath the Firestone Sill, which also yielded *Cravenoceras sp.* on Little Dun Fell.

Farther west, in the banks of Ashlock Sike [6376 3450 to 6361 3449], two sandstones, which dip westwards at 12° to 25°, with a locally worked coal cropping out between them, may correlate with the Coal Sills.

8/THE WHIN SILL

The western edge of the quartz-dolerite sill complex that underlies much of north-east England (Holmes and Harwood 1928; Dunham 1948, pp. 51–61) crops out along the Cross Fell escarpment. The sill usually consists of medium-grained dark greenish grey dolerite, and forms an almost continuous outcrop from Swindale in the south to the northern edge of the district. Over considerable distances it maintains a constant stratigraphical level, but shows a consistent tendency to rise to higher horizons northwards. The sill does not extend west of the Pennine Faults or south of the Swindale Beck Fault, suggesting that its intrusion was limited by the complex structures and stress changes on these lines.

Between Swindale and Scordale, the sill is intruded into the Lower Little Limestone, but on the south side of Scordale there is an abrupt change to a lower horizon, between the Robinson and Birkdale limestones. In a particularly good exposure at Mason Holes [756 224], on the north side of Scordale, the sill, 3 to 5 m thick, is seen to have transgressed a channel sandstone (Fig. 23). North of Gasdale, the intrusion again changes horizon, reappearing in the strata closely underlying the Tynebottom Limestone, at which level it continues northwards for over 8 km, reaching its maximum thickness for the area of about 30 m at the head of High Cup Gill. Around the head of this valley a conspicuous line of crags shows its characteristic columnar jointing. The top surface of the sill, with chilled margin, is well exposed east of High Cup Nick.

North of Knock Ore Gill the sill gradually rises through the Alternating Beds to about the level of the Scar Limestone. It is well exposed in Crowdundle and Ardale becks, and forms wide outcrops on either side of the Great Sulphur Vein. In this area, its thickness is about 20 m.

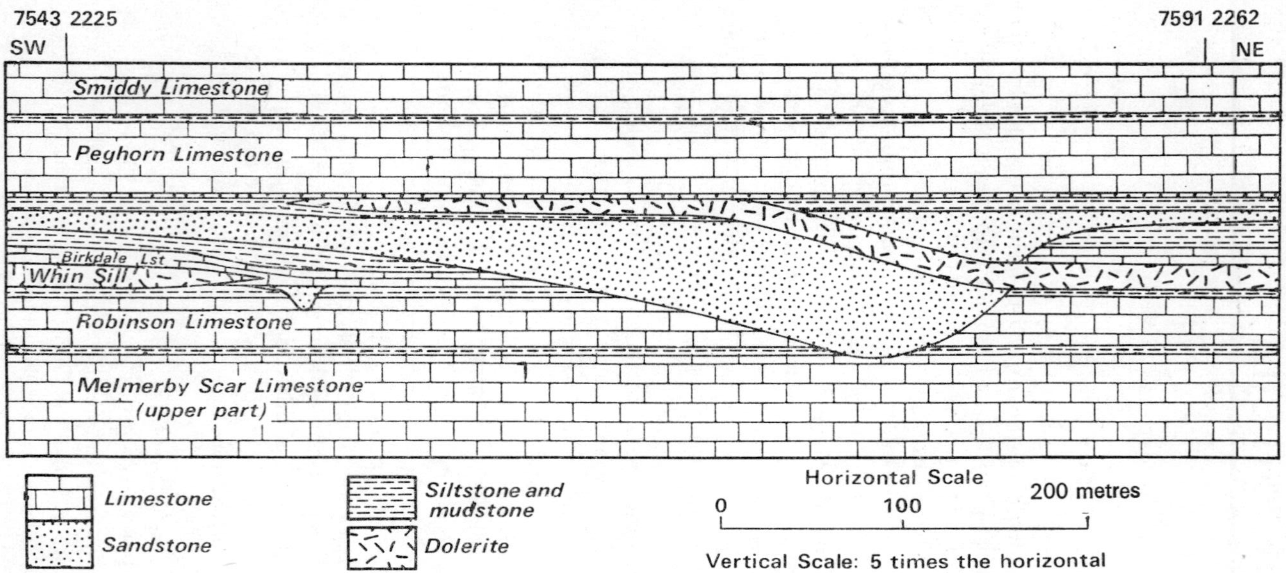

Fig. 23. *Section of Whin Sill at Mason Holes, Scordale*

Dolerite dykes related to the Whin Sill are seen on Long Fell, where they intrude Lower Carboniferous and Lower Palaeozoic strata. Olivine-dolerite dykes cutting rocks of the Borrowdale Volcanic Group at Deep Slack, Melmerby [623 381] are also of Whin Sill age.

The Whin Sill intrudes Carboniferous rocks, and pebbles probably derived from it are found in the Lower Permian brockrams of the Vale of Eden. This evidence of a late Carboniferous age is confirmed by radiometric measurements which give ages of about 295 ± 6 m.y. for the Whin Sill (Fitch and Miller 1967) and 296 ± 8 m.y. for the dykes at Deep Slack (Wadge and others 1972).

9/PERMIAN AND TRIASSIC

Permian and Triassic rocks are extensively covered by drift over much of the lower ground. They are folded into a syncline, trending north-west, roughly parallel with the line of the Pennine Fault-zone and closing only a short distance to the south of the district. They rest with marked unconformity on Carboniferous strata.

The rocks are divided into three lithological units, the Penrith Sandstone at the base, overlain in turn by the Eden Shales and the St Bees Sandstone. Of these only the Eden Shales has provided direct evidence of age; plants from near the base of that formation, and rare marine fossils found in the Belah Dolomite are Upper Permian forms. By analogy with similar deposits elsewhere in north-west Europe the Penrith Sandstone is likely to be of Lower Permian age, while the St Bees Sandstone may be, at least in part, Triassic.

Penrith Sandstone

The Penrith Sandstone was laid down under desert conditions in a basin probably coinciding approximately with the present Vale of Eden. It is a coarse-grained, millet-seed, aeolian sandstone, commonly strongly cross-bedded, and believed to be an accumulation of ancient sand dunes. Towards the contemporaneous basin margins, where relief was considerable, the aeolian sandstones are interbedded with water-laid evenly bedded sandstones, with bands of coarse breccia which are known locally as brockram and are presumed to be the product of flash-floods and wadi erosion. Within the present district, the clasts are mainly of Carboniferous limestone and sandstone and appear to be largely derived from the eastern margin of the depositional basin, an escarpment formed by early Permian down-west faulting along the Pennine Line.

The Penrith Sandstone crops out near Hilton Beck and Flakebridge on the western margin of the syncline. Outcrops

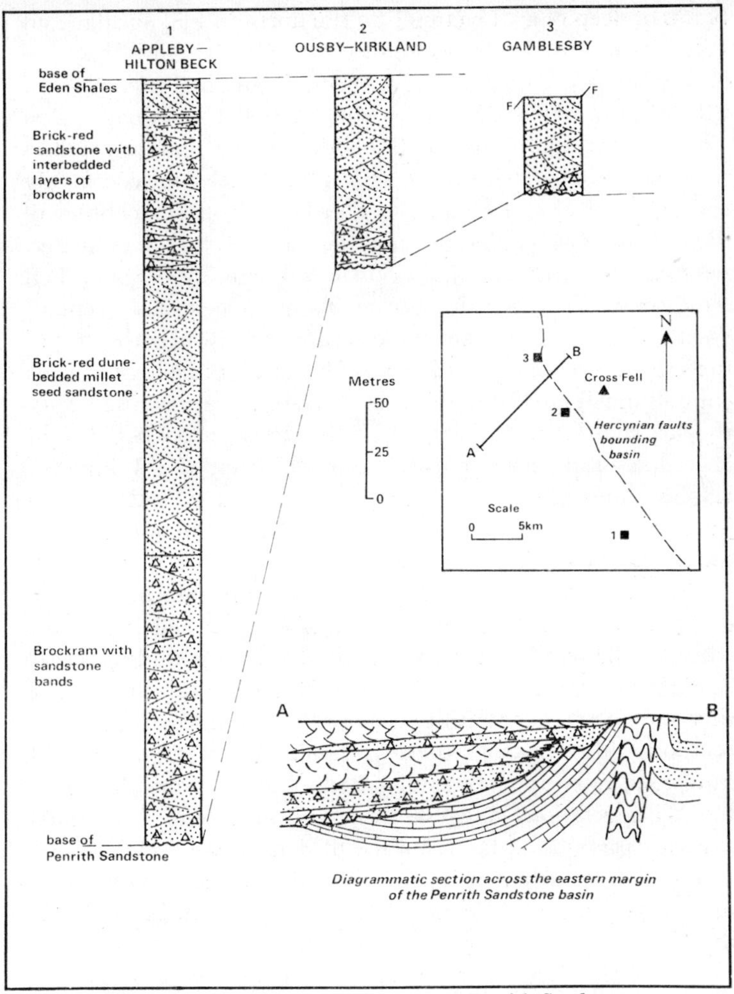

Fig. 24. *Comparative sections in the Penrith Sandstone*

indicate that the formation is some 400 m thick, and falls into three broad divisions (Fig. 24). The basal 150 m, exposed in several quarries west of the sheet boundary, are coarse brockram [Hungriggs Quarry, 6900 2125], mainly composed of dolomitized Carboniferous limestone clasts with some sandstones, set in a matrix of a millet-seed sandstone cemented with calcite. Overlying the brockram are about 150 m of poorly cemented red, dune-bedded, millet-seed sandstone, best exposed in Hilton Beck, but also seen in small sections in the banks of deep glacial channels to the north in Flakebridge and Rheabower Woods. The topmost 100 m of beds are well exposed in mural sections in Hilton Beck beginning 250 m N of Ellerhome and continuing to the base of the Eden Shales. They consist mainly of evenly bedded, water-laid, red millet-seed sandstones, with thin lenses of brockram commonly resting in erosional channels. The latter contain cobbles of dolomitized Carboniferous limestone, up to 30 cm in diameter, pebbles of Carboniferous sandstone, including Roman Fell Sandstone, and sporadic Lower Palaeozoic rocks (Kendall 1902). Dune-bedded sandstones reappear towards the top of the section. In George Gill, west of Espland, these highest beds contain much more dune sand and less brockram. This locality is particularly notable for the occurrence of pebbles of decomposed dolerite, possibly derived from the Whin Sill (Holmes and Harwood 1928; Dunham 1932). Farther north in Murton and Keisley becks, the brockrams thin out and only dune-bedded sandstone is present.

On the eastern limb of the syncline, the unconformable base of the formation is exposed south of Roman Fell [757 190], where millet-seed sandstone with brockram bands overlies reddened Carboniferous (probably Namurian) sandstones and siltstones. The unconformity is again seen in a stream north of Kirkland [6431 3362], where a small outlier of millet-seed sandstone rests upon Basement Beds, but at several localities the field relations suggest that the sandstone is banked eastwards against a land surface of considerable local relief. This is clearly displayed between Windy Hall and Wythwaite, particularly north of Windy Hall [6589 3045], in Crowdundle Beck [6600 3134 to 6586 3138] and in Littledale Beck [6605 3167], where the aeolian sandstone appears banked against reddened dolomitic limestone.

Farther north, the slope of Carboniferous rocks on Bank Rigg rising steeply eastwards from the Penrith Sandstone outcrop, may be a relic of the early Permian land surface. The relationship is best seen in Kirkland Beck [6515 3262], where a small pocket of millet-seed sandstone rests in a hollow on that surface. The land surface appears to extend as far north as Fellside Farm, with good sections showing reddened Carboniferous beds closely overlain by aeolian sandstones, exposed in Ardale Beck [6359 3410 to 6377 3401] and Ashlock Sike [6338 3438 to

6362 3450], and still farther north, a similar section [6245 3896] is seen south-east of Gamblesby.

Local bands of brockram are best seen in Crowdundle Beck, where wind-polished clasts of limestone and sandstone seem to derive from the nearby Carboniferous outcrops, and near Ranbeck Farm [6551 3216] where the pebbles include Lower Palaeozoic siltstones, greywackes and volcanic rocks. A hard brockram of limestone and sandstone clasts set in a calcareous sandstone matrix lies near the base of the formation near Gamblesby [6235 3888]. Water-laid beds are best exposed near Fellside Farm [6327 3505], where thin partings of pale grey, fine-grained sandstone occur within the aeolian facies.

Exposed thicknesses of the formation are difficult to assess accurately, because the structural dip is masked by dune-bedding, but it is calculated that about 100 m of sandstone are present between Ousby and Kirkland, compared with only about 50 m towards Milburn, and in the north near Gamblesby.

Eden Shales

Over much of the Vale of Eden, the desert sandstones of the Penrith Sandstone are succeeded by a continental sabkha facies of grey and purplish red, thinly bedded sandstones and siltstones, subordinate grey mudstones with plant remains, and beds of dolomite and gypsum-anhydrite (Fig. 25, A, B, C, D). Rather more than 100 m of these sediments extend up to the Belah Dolomite which marks a marine incursion. Their formation may well span the entire period of deposition of the Magnesian Limestone of Durham and east Yorkshire, for the Belah Dolomite is tentatively equated with the Upper Magnesian Limestone (Burgess 1965). The succeeding beds revert to a continental sabkha facies but were deposited under strongly oxidizing conditions, so that their predominant colour is brick-red. Plant remains are absent and millet-seed sand grains appear, possibly derived from outcrops of Penrith Sandstone marginal to the playa. The top of this upper division, which totals about 80 m, is taken at the first appearance of massive sandstones. This boundary is almost certainly diachronous, but affords the best lithostratigraphical distinction between the Eden Shales and the overlying St Bees Sandstone.

The complete sequence of the Eden Shales has been proved in a borehole near Hilton [7284 2056] (Fig. 25) (Burgess and Holliday 1974), but the only good surface section is in Hilton Beck, between 1300 and 600 m W of Hilton Bridge [7194 2055].

The lowest beds of the Eden Shales in the Hilton Borehole are the Hilton Plant Beds, which are thinly bedded, yellow,

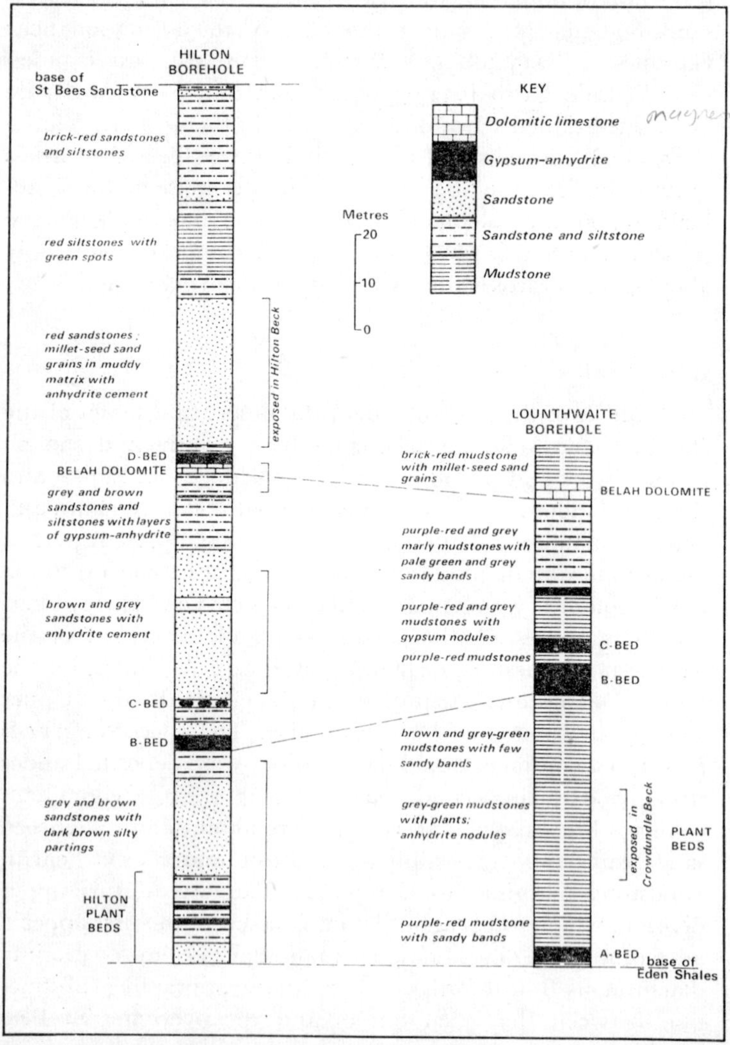

Fig. 25. *Comparative sections in the Eden Shales*

dolomitic sandstones interbedded with grey and black siltstones crammed with fragmentary plant remains with several thin beds of gypsum-anhydrite. The plant beds are exposed in the beck [7190 2063] where the commonest fossils are *Lepidopteris martinsi*, *Pseudovoltzia liebeana*, *Ullmannia bronni* and *U. frumentaria*. These beds are believed to be the lateral equivalent of the A-Bed gypsum-anhydrite, which is worked for gypsum a few kilometres to the north-west at Birks Head. They pass upwards into hard, thinly bedded, reddish brown sandstones with intervening dark brown shales which are exposed in the beck up to a small footbridge [7202 2069]. The horizons of the B-Bed and C-Bed gypsum-anhydrite deposits are not seen at the surface, probably because of solution, although succeeding strata up to the Belah Dolomite are exposed at intervals. In the Hilton Borehole, the sandstones in this interval were cemented almost exclusively with anhydrite, which dissolves away near the surface leaving very soft and easily eroded sand. The dolomite is well exposed in the stream, and on the south bank where it is partly involved in a landslip [7231 2050]. The D-Bed gypsum-anhydrite, which immediately overlies it, is absent at outcrop, due to solution. A thin solution residue and a collapse breccia mark its former position. The succeeding strata are brick-red sandstones with green reduction-spots, containing abundant millet-seed sand grains in a red mud matrix. Again the borehole cores show the cement to be largely anhydrite, and the beds are poorly exposed at the surface. The topmost 1 m, however, has a carbonate cement and projects as a hard rib [7239 2045]. It is overlain by brick-red thinly bedded, ripple-marked sandstones and siltstones that crop out just above the bend on the south bank. Beyond this, and up to the base of the St Bees Sandstone, there is a gap in the section corresponding to the outcrop of the soft marls and sandy siltstones seen in the upper part of the borehole.

Other sections in the south are fragmentary. The basal beds of the sequence are seen near Espland [7228 1893], and in Murton [7161 2157] and Keisley becks [7020 2293]. On the east side of the syncline, short sections occur in a stream south-west of Roman Fell [7268 1880] and in Milburn Beck [6746 2827].

Farther north, all except the uppermost part of the sequence was proved in a borehole at Lounthwaite [6535 3092]. The section is similar to the Hilton sequence (Fig. 25), although it contains thicker beds of gypsum-anhydrite, both in the lowest strata, of Hilton Plant Bed facies, and in the succeeding mudstones.

Plant-bearing grey mudstones, lying near the base of the formation, are exposed in the adjacent stream-section in Crowdundle Beck [6561 3133]; the commonest plants here are *Lepidopteris martinsi*, *Pseudovoltzia liebeana*, *Strobilites bronni*, *Ullmannia bronni* and *U. frumentaria*. Brick-red mudstones containing much millet-seed sand crop out farther downstream [6465 3061]. The latter probably lie above the Belah Dolomite, a view supported by the presence here of a hard calcareous band similar to that recorded at this level in Hilton Beck.

South-east of Gamblesby [6159 3872], a brief section of reddish brown mudstones with sandy bands appears to lie at the top of the Eden Shales.

The gypsum-anhydrite deposits of the Eden Shales are a valuable commercial source of gypsum; A-Bed and, more extensively, B-Bed have been exploited by both quarrying and mining (Hollingworth 1942; Sherlock and Hollingworth 1938). Within the Cross Fell district, the only active mine is at Birks Head [6678 2574] west of Dufton, on the present southern limit of the productive area.

St Bees Sandstone

The St Bees Sandstone is part of an extensive sheet of water-laid sandstone that covered northern England at this period, the base of which conventionally has been taken to mark the start of the Triassic system. Consisting mainly of even- or cross-bedded sandstones in units 30 to 60 cm thick, with silty partings, it varies in colour from brick-red to yellowish or white and, in contrast to the Penrith Sandstone, contains both feldspar and mica. The rock is generally well cemented, and the number of large quarries to be seen attests to its former value as a building stone.

Good exposures are visible on the banks of Hilton and Murton becks, in Dufton Gill, and in Swindale Beck, Knock. Farther north, the formation is cut by many glacial channels, such as Sunnygill Wood and Melmerby Beck. At least 250 m of the sandstone probably lie in the syncline west of Ousby.

10/STRUCTURE

The structural history of the Cross Fell district falls into two distinct parts, one before and one following the intrusion of the Weardale Granite. Throughout the Lower Palaeozoic, the district lay in an active orogenic belt; and its history was one of geosynclinal sedimentation, island-arc volcanicity and regional dynamic metamorphism. After the end of the Caledonian orogeny, following the final phase of granitic intrusion, the district lay within a major continental mass; and its later history is one of shelf-sea marine, deltaic or continental sedimentation, gentle crustal warping and block faulting.

Before the intrusion of the Weardale Granite

The different Lower Palaeozoic rock groups show contrasting styles and intensities of structure (Shotton 1935; Moseley 1972), and will be discussed in stratigraphical order.

Structures in the Skiddaw Group

These rocks have been affected by several phases of deformation. Where fully developed each phase is characterized by a set of minor folds (F_1, F_2, etc.) and an associated cleavage (S_1, S_2, etc.). The most highly deformed rocks in the inlier lie within the southern outcrop of the Murton Formation between Murton Pike and Brownber, where a sequence of phases can be tentatively established. Elsewhere there are fewer folds, and cleavages are weak or absent, so that successive phases cannot be distinguished.

The intensity of folding decreases rapidly northwards within the inlier, so that the rocks at the northern outcrop of the Murton Formation are only moderately deformed. The folding within the younger Kirkland Formation is also less intense.

In the southern outcrops of the Murton Formation, the earliest folds (F_1) are nearly isoclinal in style and are commonly asymmetrical, with short limbs of 1 to 5 cm and long limbs exceeding 30 cm. The folds are difficult to detect in mudstones, especially where the short limbs are sheared out, as in the intensely folded areas on Brownber Hill [7046 2715], on the ridge north of Keisley Beck [7215 2415] and in the upper part of Murton Beck [7377 2236]. Here the bedding generally lies almost parallel to the cleavage planes which are lustrous with secondary chlorite flakes. In siltstones the bedding is commonly picked out by pale grey sandy bands which may thicken markedly in the fold axes, so that an axial trace appears on a cleavage surface as a ridge of quartz blebs. The fold axes are generally associated with limonitized pyrite cubes up to 5 mm across.

A strong slaty cleavage (S_1) associated with these folds is the dominant surface visible in most exposures. Its orientation is variable as a result of later movements, but on Brownber Hill, where there seems to have been the least subsequent disturbance, it is vertical and trends at about 100°. The F_1 folds show a very variable plunge in the plane of the cleavage.

The second fold phase visible in the field is characterized by asymmetrical open folds (F_2) with a wavelength of about 1 to 2 m. The foliation which is axial planar to these folds is generally a fracture cleavage (S_2), variably developed and commonly appearing as a strong crenulation on S_1. Around Brownber Hill [7046 2715], both S_2 and the F_2 axial planes are almost horizontal, which is taken to be close to their original orientation. Elsewhere, S_2 may dip at any angle up to vertical, as a result of later movements. Most of these later movements appear to have occurred about axes parallel to the Pennine Line, since both the plunge of the F_2 folds and the trend of S_2 are commonly aligned at 325° to 335°; and they may be of a much younger age. Much of the quartz-veining in the Murton Formation, as on Brownber [7062 2744], seems to date from the F_2 phase.

A third cleavage (S_3) is visible on Brownber Hill. It is aligned vertically, trends east–west and is axial planar to very open, gentle F_3 folds seen affecting S_2. The S_3 trace on S_2 is generally close to horizontal.

In addition, the Murton Formation is affected in the southern outcrops by narrow zones, 1 to 5 m wide, of very intense folding which post-date all the previous minor structures. The zones generally dip steeply eastwards, and consist of folds which are tight but with no axial plane cleavage. The folds generally trend at 340° to 350°, but their plunges appear to be completely random. The zones are best seen in Swindale Beck [6947 2854] and Murton Beck [7378 2233]. They commonly lie close to low-angle reverse faults of post-Carboniferous age and may result from the same stresses.

In the outcrops of the Murton Formation at the northern end of the inlier, large-scale folds with a Caledonoid trend can be readily detected (Fig. 15), but are difficult to correlate with the structures farther south. It is tentatively suggested that they equate with the F_1 structures on Brownber but a correlation with F_3 is equally possible. The folding is indicated mainly by the dips of the bedding, which is the dominant foliation hereabouts. A very weak cleavage, best seen around Kits How [6405 3549], lies axial planar to the large folds which have scattered minor folds on their limbs. The style and orientation of these minor flexures give some indication of the geometry of the major structures. For example, near the axial region of the syncline on Catterpallot Hill [638 363], minor folds are relatively numerous and suggest that this major structure plunges generally eastwards and that its axial plane dips steeply to the north.

Later folding, trending north-north-westerly and very open in style, is present in a few outcrops. It is best seen on Baron Side [6403 3566] and in Dale Beck [6364 3592], where it is accompanied by a rare axial planar fracture cleavage.

Folding predating the deformations described above may also be present but the evidence is limited. A feeble bedding schistosity is developed in places, as for example in Dry Sike [6387 3732]. Elsewhere, narrow belts of complex, apparently disorientated minor folds, which lack any cleavages, lie along the general Caledonoid strike. They are best seen in Dry Sike [6402 3738].

The adjacent outcrops of the Kirkland Formation show fewer cleavages or fold closures, although dips are generally steep. In the south of the inlier, the beds exposed east of Murton show in places a weak near-vertical cleavage trending east-north-east. Farther north, between Flagdaw and Ardale Beck, the minor structures in the Kirkland Formation outcrops appear to correspond with the F_2 and F_3 structures in the Murton Formation, although this correlation cannot be proved. The possible F_2 phase structures are flat-lying open folds trending north-west and generally associated with a weak fracture cleavage. They are best seen in Eller Gill [6757 3121] and in Milburn Beck [6787 2925]. Much more common however are the possible F_3 correlatives, which are open folds of east-north-east (Caledonoid) trend. Within the mudstones of the Kirkland sequence, these folds are comparatively small and tight with a weak axial plane cleavage. Typical examples can be seen in Eller Gill [6763 3118] and Ardale Beck [6469 3434]. Where the competent volcanics are involved, however, the folds tend to be broad and open; these are best seen on Wythwaite Top [665 324] but are also present on Flagdaw, the west side of Burney Hill and in Mudgill Sike [6737 3054].

Structures in the Borrowdale Volcanic Group

Direct evidence of the relationship between the folding in the Skiddaw Group and that in the Borrowdale Volcanic Group is lacking in the inlier as all the contacts are faulted. By analogy with the Lake District (Simpson 1967), at least part of the polyphase deformation in the Skiddaw Group probably predates the deposition of the Borrowdale Volcanic Group.

The structures in the Borrowdale Volcanic Group are much less complex than those in the Skiddaw Group. The massive volcanic rocks are generally strongly jointed but lack a good cleavage. Softer tuff bands on Harthwaite and Dufton Pike are cut by a weak cleavage, striking like the bedding at about 160° and dipping steeply to the south-west. This cleavage may be associated with north-trending pre-Caradoc folds like those recorded in the Lake District outcrops (Moseley 1972). The volcanic rocks are also affected by open folds trending between east and north-east dating from the end of the Silurian. The outcrops on Knock Pike, Dufton Pike, Keisley and Roman Fell appear to mark anticlines of this type.

Structures in the upper Ordovician and Silurian sediments

As a result of the Caledonian (end-Silurian) folding these rocks are thrown into open folds with a near east–west trend, and most exposures also show some cleavage striking between 75° and 105°. Minor folds with this orientation are well exposed in Swindale Beck, Knock, where the beds are inclined generally southwards and zones of steep dip (60° to 80°) alternate with zones of low dip (20°). The cleavage is well developed in the shaly beds, dipping southwards at 65° to 85°; in the more massive lithologies the cleavage is more obvious where dips are low, since the foliation tends to coincide with the bedding planes on the steep fold limbs.

After the intrusion of the Weardale Granite

The Pennine Line developed, shortly after the intrusion of the Weardale Granite, along the western edge of the Alston Block and included the Fellside, Knock Pike, Dufton Pike and Swindale Beck faults. The large down-westerly throw on these fractures preserved younger Lower Palaeozoic rocks to their west whilst the upthrow side was so deeply eroded that only Skiddaw Group beds underlie the Carboniferous rocks on the western edge of the Alston Block. Many of the smaller faults in the inlier probably date from this time also. This can be well demonstrated below Roman Fell where several fractures pass beneath undisturbed Carboniferous sediments.

During Carboniferous times, the Alston Block remained an area of comparative uplift, in contrast to the Northumberland and Stainmore troughs lying respectively to the north and south. The block was not finally submerged beneath the sea until Viséan times and was subsequently covered by thinner sequences of sediment than those of the adjacent troughs. Differential subsidence persisted throughout the Namurian, as the block sequence is thinner than that in the Stainmore Trough (Owens and Burgess 1965) or in the Vale of Eden.

The Armorican earth movements at the end of the Carboniferous locally took the form of an east–west compression. Reaction to the stress was concentrated along the zone of weakness marked by the Pennine Line. Here, the brittle rocks of the basement broke along several low-angle reverse faults

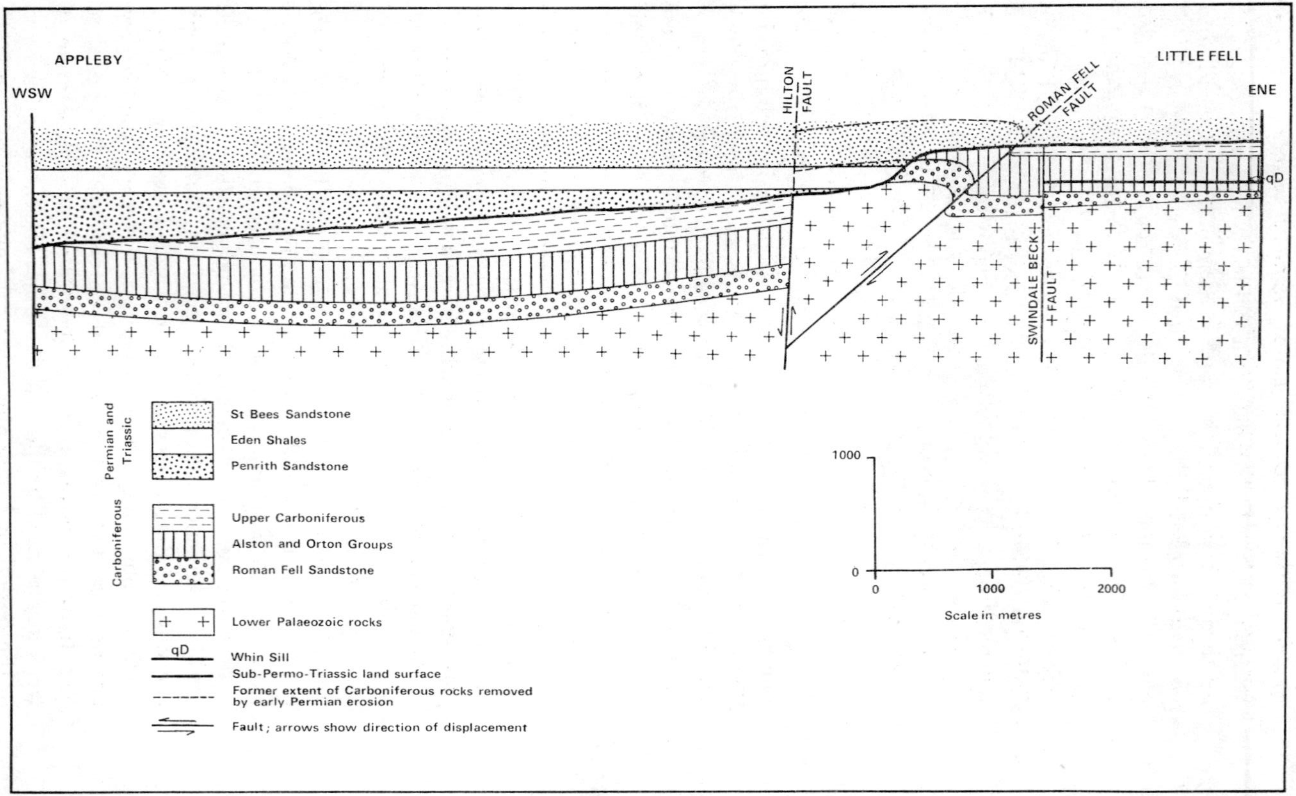

Fig. 26. *Diagrammatic section across the Pennine Line in early Triassic times*

and were pushed eastwards over adjacent parts of the block (Brownber Fault, Murton Pike Fault). In places these movements occurred along pre-existing fractures initiated during the Caledonian faulting. The more plastic cover of Carboniferous rocks responded initially by tight folding along axes parallel to the underlying faults, and then by the development of an eastwards-facing monocline with nearly vertical or overturned beds on the steep limb. Where the compression was greatest the reverse faults penetrated into the Carboniferous cover. The overall displacement in the south, between Roman Fell and Brownber, was at least 250 m (Fig. 26) and in the north may have been much greater (Wadge and others 1972).

The Murton Pike, Brownber, Lad Slack and Deep Slack faults show the deeper levels of these structures where, in each case, most of the movement has occurred along a well-defined fracture. The higher structural levels can be examined on the Roman Fell Fault and on subsidiary dislocations, such as the Great Sulphur Vein, which have behaved in a similar way to the Pennine Line. It is likely that lateral movement occurred on these faults, but this is difficult to prove generally. The folds on Roman Fell end abruptly against the Swindale Beck Fault and may indicate sinistral tear movements along it.

As a result of the compression, the Carboniferous rocks of the Vale of Eden were thrown into a broad syncline, with dips of up to 10°, which plunged gently to the north-west. The uplifted eastern margin of this fold, along the Pennine Line, was deeply eroded, exposing sandstones and siltstones, probably of Namurian age, in the south (Roman Fell) and limestones and sandstones of Lower Carboniferous age in the ground between Milburn and Ousby.

The intrusion of the Whin Sill and its associated dykes appears to have closely followed this period of compression (Dunham 1948, p. 65).

As the compressive stresses died away, subsidence of the Vale of Eden and down-west faulting along the Pennine Line combined to form a cuvette in which Permian sediments began to accumulate. Locally the movements occurred along both pre-existing fractures and new but sub-parallel faults close by. The nature of this double movement of the Armorican faults is well illustrated by the position of a fault-bounded slice of Melmerby Scar Limestone [718 246] north-east of Keisley.

It lies along the reverse down-east Brownber Fault, but must also be bounded by a fracture throwing at least 250 m down-west from the limestone outcrop on Peeping Hill.

A remnant of the westwards-facing fault scarp formed by these movements is preserved on Roman Fell, where both Lower Palaeozoic rocks and Roman Fell Sandstone are deeply reddened on the western face of the fell, but not farther east (Burgess and Harrison 1967). Comparable reddening of Lower Palaeozoic rocks is also seen farther north at Keisley (Keisley Limestone and Brathay Flags) and east of Milburn (Knock Pike Tuffs). The preservation of Namurian strata on the sub-Permian cuvette floor near Roman Fell, at a lower topographic level than the reddened Roman Fell Sandstone just east of the Hilton Fault, implies that the early Permian down-west throw was there of the order of 800 m (Fig. 26). The scarp may not have been finally buried till Triassic times (St Bees Sandstone) as a band of brockram containing presumed Lower Palaeozoic clasts is present in the highest beds of the Eden Shales near Dufton [6820 2486] (Versey 1939).

East of the inlier, reddening of Carboniferous rocks has been noted at only one locality, on the summit of Little Fell, where it affects Namurian sandstones. If this alteration is of Permian age, a further eastwards extension of the Permian land surface may be inferred (Fig. 26).

The latest movements affecting the area are ascribed to the Alpine orogeny but are difficult to date accurately. Renewed uplift and easterly tilting of the Alston Block was accompanied by down-west faulting along many of the earlier fractures. A total displacement of about 300 m around Roman Fell and about 600 m near Cross Fell is estimated. The present Pennine escarpment is the degraded fault-line scarp produced by these movements which probably continue to the present day. An earthquake of unusual intensity for this country occurred near the southern end of the Pennine Line (Browning and Jacob 1970) in August 1970.

Also as a result of these stresses, the Permian rocks of the Vale of Eden were folded into an asymmetrical syncline on an axis parallel to the Pennine Line, with a gentle dip on the western limb, but with dips of up to 40° on the east.

11/MINERALIZATION

The Cross Fell area lies on the western edge of the Northern Pennine Orefield and the Carboniferous rocks of the escarpment are cut by mineral veins at numerous localities (Dunham 1948, pp. 123–41). The veins occur along faults, generally with an east–west or east-north-east trend. The oreshoots take two forms, occurring either as lenses following the faults (vein oreshoots) or as replacements of the thicker limestones on either side of the faults (metasomatic flats). Most of the veins were extensively investigated in the past, principally in the search for galena (lead ore). Exploratory levels, some of which are still accessible, were driven at many localities. They are commonly in a highly dangerous condition, and should not be entered without expert guidance. Surface exploration was principally by means of 'hushing'—a process whereby ponded water was suddenly released and channelled either along known veins or across the possible course of veins—resulting in the rapid removal of surface deposits and erosion of the solid rock along the chosen line. Deep gullies formed by this process are common, the most impressive example being Dun Fell Hush [715 319], over 500 m long and up to 15 m deep.

Galena usually forms only a small part of the vein material, the remainder (gangue) being either barite or fluorite. These minerals, discarded by the early miners, are now of commercial value, the former for chemical purposes, for drilling mud and for use as an aggregate for concrete used in radiation shielding, the latter also for chemical purposes and for use as a flux in the steel industry. Barite is found in small quantities in the mines on Long Fell, Murton Fell and Dufton Fell. The principal deposits are on Great Dun Fell, where veins and flats occur about the horizon of the Great Limestone. Silverband Mine [709 319] provided the main access to these deposits and considerable reserves of barite are believed still to be present in the area. Fluorite is not present in commercial quantities, though it does occur in Scordale, where good specimens of amber fluorite may bc obtained from the old mine tips at the head of the valley.

Limonite (iron ore) is found in the veins east of Dun Fell and, farther north, at Ardale Head, but the deposits have not been exploited commercially.

North-east of Melmerby, the Great Sulphur Vein carries a different type of mineralization. The gangue mineral is quartz, and the associated ore minerals include pyrrhotite, pyrite, marcasite and subordinate chalcopyrite.

The mineralization post-dates the intrusion of the Whin Sill, and is believed, on evidence from other areas, to be most probably of Upper Permian or slightly younger age.

12/SUPERFICIAL DEPOSITS AND ASSOCIATED LANDFORMS

A till-sheet covers most of the lower ground of the district and consists mainly of boulder clay, although silts, sands and gravels are intercalated within the thicker parts of the sheet. The spread thins upslope to the Pennines and is rarely present above 1500 ft (457 m) OD. Numerous glacial channels were cut by meltwater across the lower ground and are associated locally with extensive deposits of sand and gravel.

In contrast, on the higher ground of the Pennines, the commonest superficial deposit is head, formed by solifluction under periglacial conditions, and commonly eroded to give boulder fields. Many solid outcrops on the fells are cambered and on the steeper slopes landslips are common.

Trotter (1929) and Hollingworth (1931) postulated several glaciations, each marked by its own till-sheet, and with interbedded sands and gravels thought to represent inter-glacial periods. The existence of such warmer periods can only be established by the occurrence of organic layers interbedded with the glacial sequence. The only such bed recorded lies beneath boulder clay in Scandal Beck at the southern end of the Vale of Eden, and has yielded a radiocarbon date of 34 500 years BC (Shotton and others 1970). Accordingly, a late Devensian age is assigned to the succeeding glacial deposits in Scandal Beck and, by analogy, to the similar deposits of the Cross Fell area. This age is supported by the freshness of the landforms produced during the deglaciation of the present district.

Boulder clay

The boulder clay consists of rock fragments ranging in size from boulder to sand grade set in a matrix of silty clay or clayey sand. It is generally stiff and well consolidated, in contrast to the morainic mounds of poorly consolidated englacial or supraglacial debris deposited in the larger Pennine valleys and high on the escarpment. On the low ground the erratic boulders in the till are principally of Lake District provenance, including the conspicuous and easily identifiable Shap Granite. Closer to the escarpment, however, a locally derived till characterized mainly by Skiddaw Group mudstone, Carboniferous sandstone and Whin Sill erratics is present. The matrix of the till generally reflects the lithology of the bedrock. Over the inlier, it is yellowish grey or brown clay, whereas on the Permo-Triassic rocks it is reddish brown and very sandy, appearing in places to grade down into the underlying bedrock. An exception to this is seen in the Robberby area [590 366] where a stiff, grey boulder clay of Lake District provenance containing many large Borrowdale Volcanic Group erratics (predominantly andesite) rests on the St Bees Sandstone. Over most of the low ground the till sheet is seldom more than 5 m thick and in some areas, as in the deeply channelled ground between Ousby and Skirwith, it has been extensively eroded. Elsewhere, notably in the area east of Blencarn, it thickens to form drumlins.

The till was mainly laid down by an ice-sheet which spread eastwards into the Vale of Eden from the Lake District and then divided into two streams, one continuing eastwards across the Stainmore col into Teesdale, the other moving north-westwards down the Vale. South of Knock, the direction of ice-movement along the escarpment appears to have been south-easterly—as boulders of the Dufton Microgranite [692 268] are found only in this direction from the outcrop—swinging to easterly over Long Fell where erratics of Roman Fell Sandstone are widely distributed. North of Knock the alignment of drumlins suggests a north-westerly flow.

Though local glaciers issuing from the larger Pennine valleys contributed to the ice-sheet, there is no evidence to suggest that there was a major ice-cap on the Cross Fell ridge; it seems probable, also, that the high ground between Little Fell and Cross Fell was not overridden at any time by Lake District ice, which did not rise much over 2000 ft (610 m) OD on the escarpment. This is supported by the absence of high-level glacial channels (see below), the presence of a thick mantle of head, and the intense cambering of strata on these fells, contrasting with the cleanly scraped outcrops east of Roman Fell, where the ice is known to have overridden the scarp. A small moraine [750 267] east of High Cup Gill marks the termination of a local glacier in the headwaters of Maize Beck.

Glacial channels

Glacial drainage channels are common throughout the Vale of Eden and on the lower Pennine slopes. They are believed to result largely from the sub-glacial flow of meltwater, a view contrasting with earlier ideas (Trotter 1929; Hollingworth 1931) that they were cut by meltwater flow marginal to a wasting ice-sheet.

Channels often contain little or no drainage at the present day, and may begin or end abruptly. They are cut impartially into hard or soft rocks, and their course usually bears little relation to the regional slope of the ground; individual channels may follow, and locally even ascend, the topographic contours. They range greatly in size, being 100 m to several kilometres long, 15 to 200 m wide, and 2 to 30 m deep. Typical cross-sections of the smaller channels are U-shaped, whilst longitudinal profiles generally slope smoothly northwestwards, in the direction of flow under gravity, although in places the profile is humped, indicating the uphill flow of confined meltwater under hydrostatic pressure. The larger channels, on the low ground, may be very wide and flat-bottomed.

Channels are present at a height of 1600 ft (488 m) OD on Melmerby Low Scar and almost 2000 ft (610 m) OD on Kirkland Fell, but are absent from all the Pennine summits, supporting the suggestion that the water-shed was covered only by thin ice or snow, incapable of producing large quantities of meltwater. The meltwater in many of the channels high on the escarpment flowed mainly along the slopes and may have been marginal to the ice-sheet in places, before running steeply downslope and cutting chutes which end abruptly where the water flowed back into the ice. Good examples of this type of channel are seen below Melmerby Low Scar [6313 3783 to 6277 3827]. Wildboar Nook [6672 3331 to 6667 3355] and on Brownber Hill [705 273].

Meltwater within the ice was ponded back to different levels in each of the Pennine valleys, and flowed from high to low water-table areas cutting channels wherever it intersected the land surface. Such channels are common on the flanks of the spurs on the Pennine scarp. They are well developed on Murton Pike [730 225], and on Baron Side, north-east of Ousby where small chutes [6403 3567] dropping into plunge-pools [6397 3562] can be distinguished.

One of the largest and most spectacular channels in the district has a humped longitudinal profile with a col below Cocklock Scar [6535 3398], whilst farther south, several large channels are incised into the lower ground between Crowdundle Beck and Burney Hill, and in the area north of Keisley.

Complex systems of channels, dendritic in plan, with many tributaries flowing at grade into a major valley, are common on the lower ground. The sub-parallel orientation of the tributary channels lying about 200 to 400 m apart was probably controlled by lines of weakness such as major crevasses or shear-surfaces in the ice.

In the area south of Knock, all the low-ground channels, from the extreme south-western edge of the district as far north as Swindale Beck seem at some time to have formed parts of a unified sub-glacial drainage system which carried vast quantities of meltwater into the wide flat-bottomed channel running north-westwards from Langtonfield [709 203] by Stock Bridge [686 235] to Long Marton, beyond the south-western edge of the map. Farther north, other dendritic systems are developed, particularly near Ousby.

In summary it appears likely that the channels may belong to two systems, an early high-level system running along and within the ice margin, and a later more comprehensive sub-glacial system draining towards the centre of the valley as the stagnant ice-sheet rapidly wasted away. The complex form of this later drainage pattern precludes the possibility of individual channels marking the positions of wasting ice-margins.

Glacial sand and gravel

The sand and gravel forms eskers, irregular mounds and flat spreads. The deposits are restricted to ground at the foot of the Pennine scarp below 1200 ft (366 m) OD and to the lower ground farther west. Many of the more accessible deposits have been worked locally for aggregate, although near the inlier the gravel is rarely used even as fill, because the proportion of easily degradable clasts of Skiddaw mudstone is high.

The eskers are steep-sided, sharp-crested ridges composed largely of gravel and cobbles, with subordinate sand and silt. The best example lies in the extreme north at Whinny Hill [625 392]. Although arcuate or sinuous in plan, the eskers generally trend north-west like the glacial channels with which they are often closely associated, as for example near Fellside [638 352] and Ousby Townhead [637 336]. It is considered that these deposits were laid down as the bed-load of sub-glacial streams and subsequently left standing as ridges when the ice melted. Elsewhere, however, some eskers appear unrelated to channel drainage and probably reflect meltwater drainage within, rather than below, the ice-sheet; as at Kirk Hill near Blencarn [643 318], north of Bank Wood [642 339] and east of Kirkland [655 327].

The irregular mounds of sand and gravel lack the ridge form of the eskers, but are similar in composition, with sand generally subordinate to gravel. The mounds commonly are present on the north sides of spurs extending from the escarpment, as on Roman Fell [745 209], Middle Tongue [725 238], Dufton Pike [704 268 to 688 271], Flagdaw [690 295] and Rusby Hill [647 348], and are apparently related to marginal or submarginal glacial drainage systems, though not necessarily to specific channels in bedrock.

Several deposits have a delta-like form with flattish tops and steep downslope faces as around Dufton Pike, in the partly-worked spread north of Knock Pike [685 288] and in the 20 m-thick mound south of Crowdundle Beck [674 315]. The cross-bedding and graded-bedding in these mounds, best seen in the gravel pit on Knock Pike, indicate that they were deposited in standing water, probably ponded by the wasting ice-sheet. Other mounds may owe their irregular form to deposition upon melting ice.

Sand and gravel, possibly of rather later age, occur in places as spreads marginal to large glacial channels, notably along Hilton Beck [711 202] and on the south bank of Crowdundle Beck, north of Loscars [636 301], where up to 10 m of gravel are present.

Peat

Much of the higher ground on the Pennine escarpment is covered by blanket-peat up to 3 m thick, commonly deeply dissected. The peat consists mainly of *Eriophorum*, *Calluna* and *Sphagnum*, with a layer of birch roots, the 'forest layer', at the base. Its formation began in Boreal times (about 7500 BC) and continued till about 500 BC, since when erosion has been dominant, and many areas, including the main Pennine summits, have been stripped almost bare (Johnson and Dunham 1963, pp. 131–51). On the lower ground also, isolated areas of peat are present, commonly in hollows in the floors of glacial channels.

REFERENCES

Bott, M. H. P. 1967. Geophysical investigations of the northern Pennine basement rocks. *Proc. Yorks. geol. Soc.*, **36**, 139–68.

—— 1974. The geological interpretation of a gravity survey of the English Lake District and the Vale of Eden. *Jnl geol. Soc. Lond.*, **130**, 309–31.

Browning, G. R. J. and Jacob, A. W. B. 1970. Preliminary study of the North of England earthquake of August 9, 1970. *Nature, Lond.*, **228**, 835–7.

Buckland, W. 1817. Description of an insulated group of rocks of slate and greenstone in Cumberland and Westmoreland, on the east side of Appleby, between Melmerby and Murton. *Trans. geol. Soc. Lond.*, **4**, 105–16.

Burgess, I. C. 1965. The Permo-Triassic rocks around Kirkby Stephen, Westmorland. *Proc. Yorks. geol. Soc.*, **35**, 91–101.

—— and Harrison, R. K. 1967. Carboniferous basement beds in the Roman Fell district, Westmorland. *Proc. Yorks. geol. Soc.*, **36**, 203–25.

—— and Holliday, D. W. 1974. The Permo-Triassic rocks of the Hilton Borehole, Westmorland. *Bull. geol. Surv. Gt Br.*, No. 46, 1–34.

—— Rickards, R. B. and Strachan, I. 1970. The Silurian strata of the Cross Fell area. *Bull. geol. Surv. Gt Br.*, No. 32, 167–82.

Capewell, J. G. 1956. The Carboniferous Basement Series of the Cross Fell area, Cumberland and Westmorland. *Proc. Geol. Ass., Lond.*, **66**, 213–30.

Dean, W. T. 1959. The stratigraphy of the Caradoc Series in the Cross Fell Inlier. *Proc. Yorks. geol. Soc.*, **32**, 185–227.

Dunham, K. C. 1932. Quartz-dolerite pebbles (Whin Sill type) in the Upper Brockram. *Geol. Mag.*, **69**, 425–7.

—— 1948. Geology of the northern Pennine orefield. Vol. I, Tyne to Stainmore. *Mem. geol. Surv. Gt Br.*

Eastwood, T., Hollingworth, S. E., Rose, W. C. C. and Trotter, F. M. 1968. Geology of the country around Cockermouth and Caldbeck. *Mem. geol. Surv. Gt Br.*

Fitch, F. J. and Miller, J. A. 1967. The age of the Whin Sill. *Geol. Jnl*, **5**, 233–50.

Forster, W. 1809. A treatise on a section of strata from Newcastle-on-Tyne to the mountain of Cross Fell in Cumberland; with remarks on mineral veins in general. (1st edition) Alston.

GARWOOD, E. J. 1913. The Lower Carboniferous succession in the north-west of England. *Quart. Jnl geol. Soc. Lond.* (for 1912), **68**, 449–586.

GREEN, J. F. N. 1919. The vulcanicity of the Lake District. *Proc. Geol. Ass., Lond.*, **30**, 153–82.

HELM, D. G. 1970. Stratigraphy and structure in the Black Coombe Inlier, English Lake District. *Proc. Yorks. geol. Soc.*, **38**, 105–48.

HOLLINGWORTH, S. E. 1931. The glaciation of western Edenside and adjoining areas and the drumlins of Edenside and the Solway Basin. *Quart. Jnl geol. Soc. Lond.*, **87**, 281–359.

—— 1942. The correlation of gypsum-anhydrite deposits and the associated strata in the north of England. *Proc. Geol. Ass., Lond.*, **53**, 141–51.

HOLMES, A. and HARWOOD, H. F. 1928. The age and composition of the Whin Sill and the related dykes of the north of England. *Mineralog. Mag.*, **21**, 493–542.

HUDSON, S. N. 1937. The volcanic rocks and minor intrusions of the Cross Fell Inlier, Cumberland and Westmorland. *Quart. Jnl geol. Soc. Lond.*, **93**, 368–405.

JOHNSON, G. A. L. and DUNHAM, K. C. 1963. The geology of Moor House. *Monogr. Nature Conservancy*, No. 2.

—— HODGE, B. L. and FAIRBAIRN, R. A. 1962. The base of the Namurian and of the Millstone Grit in north-eastern England. *Proc. Yorks. geol. Soc.*, **33**, 341–62.

KENDALL, P. F. 1902. On the brockrams of the Vale of Eden and the evidence they afford of an inter-Permian movement of the Pennine faults. *Geol. Mag.*, **39**, 510–3.

LISTER, T. R., BURGESS, I. C. and WADGE, A. J. 1969. Microfossils from the cleaved Skiddaw Slates of Murton Pike and Brownber (Cross Fell Inlier). *Geol. Mag.*, **106**, 97–9.

MOSELEY, F. 1972. A tectonic history of northwest England. *Quart. Jnl geol. Soc. Lond.*, **128**, 561–98.

NICHOLSON, H. A. and MARR, J. E. 1891. The Cross Fell Inlier. *Quart. Jnl geol. Soc. Lond.*, **47**, 500–12.

OWENS, B. and BURGESS, I. C. 1965. The stratigraphy and palynology of the Upper Carboniferous outlier of Stainmore, Westmorland. *Bull. geol. Surv. Gt Br.*, No. 23, 17–44.

SHERLOCK, R. L. and HOLLINGWORTH, S. E. 1938. Gypsum and anhydrite; celestine and strontianite. *Mem. geol. Surv. Gt Br.*

SHOTTON, F. W. 1935. The stratigraphy and tectonics of the Cross Fell Inlier. *Quart. Jnl geol. Soc. Lond.*, **91**, 639–704.

—— BLUNDELL, D. J. and WILLIAMS, R. E. G. 1970. Birmingham University Radiocarbon Dates IV. *Radiocarbon*, **12**, 385–99.

SIMPSON, A. 1967. The stratigraphy and tectonics of the Skiddaw Slates and the relationship of the overlying Borrowdale Volcanic Series in part of the Lake District. *Geol. Jnl*, **5**, 391–418.

Skevington, D. 1970. A lower Llanvirn graptolite fauna from the Skiddaw Slates, Westmorland. *Proc. Yorks. geol. Soc.*, **37**, 395–444.

Temple, J. T. 1968. The Lower Llandovery (Silurian) brachiopods from Keisley, Westmorland. *Palaeontogr. Soc.* [Monogr.].

Trotter, F. W. 1929. The glaciation of eastern Edenside, the Alston Block and the Carlisle Plain. *Quart. Jnl geol. Soc. Lond.*, **85**, 549–612.

Turner, J. S. 1927. The Lower Carboniferous succession in the Westmorland Pennines and the relations of the Pennine and Dent faults. *Proc. Geol. Ass., Lond.*, **38**, 339–74.

Versey, H. C. 1939. The petrography of the Permian rocks in the southern part of the Vale of Eden. *Quart. Jnl geol. Soc. Lond.*, **95**, 275–94.

Wadge, A. J. 1972. Sections through the Skiddaw–Borrowdale unconformity in eastern Lakeland. *Proc. Yorks. geol. Soc.*, **39**, 179–98.

—— Harrison, R. K. and Snelling, N. J. 1972. Olivine-dolerite intrusions near Melmerby, Cumberland, and their age-determination by the potassium-argon method. *Proc. Yorks. geol. Soc.*, **39**, 59–70.

Printed in England for Her Majesty's Stationery Office
by McCorquodale Printers Ltd., London.
HM7424 Dd. 505167 K24 9/74 McC 569

Other Institute publications dealing with this district

Books

British Regional Geology: Northern England 40p

Geology of the country around Penrith (Explanation of 1:50 000 Geological Sheet 24) in preparation

Geology of the country around Brough-under-Stainmore (Explanation of 1:50 000 Geological Sheet 31) in preparation

Geology of the Northern Pennine Orefield Vol 1: Tyne to Stainmore £1.50

Obtainable from the Government Bookshops in London (post orders to P.O Box 569, SE1 9NH), Edinburgh, Manchester, Bristol, Birmingham, Cardiff and Belfast, or through booksellers

Geological Maps

'Ten-mile' map of Great Britain, Sheet 1 (1:625 000)

	Flat	75p
	Folded	85p

1:50 000	*Flat*	75p
	Folded and cased	85p

Penrith (24) Sheet, Solid
Penrith (24) Sheet, Drift
Alston (25) Sheet, Solid and Drift
Brough-under-Stainmore (31) Solid
Brough-under-Stainmore (31) Drift

Tectonic map of Great Britain and Northern Ireland (1:1584 000) 50p

Aeromagnetic Map

'Ten-mile' map of Great Britain, Sheet 1 (1:625 000)

	Flat	£1.25
	Folded	£1.50
	Permatrace	£3.60

Obtainable from the Institute's Bookshop at the Geological Museum and Ordnance Survey Agents